The University of Disaster

The University of Disaster

Paul Virilio

Translated by Julie Rose

polity

First published as *L'Université du désastre* © Editions Galilée, 2007

This English edition © Polity Press, 2010

Reprinted 2010

Ouvrage publié avec le concours du Ministère français de la Culture – Centre national du livre

Published with the assistance of the French Ministry of Culture – National Centre for the Book

Polity Press
65 Bridge Street
Cambridge CB2 1UR, UK

Polity Press
350 Main Street
Malden, MA 02148, USA

ISBN-13: 978-0-7456-4504-9 (hardback)
ISBN-13: 978-0-7456-4505-6 (paperback)

A catalogue record for this book is available from the British Library.

Typeset in 12 on 14 pt Bembo
by Servis Filmsetting Limited, Manchester
Printed and bound by MPG Books Group, UK

For further information on Polity, visit our website: www.politybooks.com

For Jean Duvignaud

This science whose OBJECT is the SUBJECT itself, this circular science is an unhappy science.

Vladimir Jankélévitch

Contents

I

Intuition

Is the real instant present? What reality is the history of societies loaded with? The reality of centuries, years, civilisations, past generations? Doesn't the acceleration of present reality have a decisive impact on the historicity of avowed facts? Still more precisely, is a history of REAL TIME still historical? So many questions that today affect the anthropology of the time devoted to reflection.

Can we, in fact, talk about a *contemporary* world? Shouldn't we rather talk about the anthropology of a world that is not atemporal, timeless, but atemporary?

'The survival of intuition beyond the moment should be backed,' Vladimir Jankélévitch warned us. Is an anthropology of the instant even conceivable? Can it be 'logical' without denying its fully historical dimension? To my mind, these questions, with their flurry of causes, now impose themselves on us, galvanizing our intelligence.

They say the Earth is flat, don't they? Certainly, the horizon is there for all to see, confirming the relative flatness, indeed the platitude, of the real space involved in the geopolitics of nations. But is time, real time, levelled out for all that? What's happening with the levelling out of duration, the long durations of History? Does instantaneity crush all temporality, all chronological bumps, or does the reverse happen? Does 'the evidence of a time without object which is not the time involved in history' in fact definitively spirit away our whole array of memories?[1] If that is indeed the case, a UCHRONIA won't be long in supplanting UTOPIA; and the insularity of

'World Time', which is also astronomical, will radically over-
haul Thomas More's Utopia from top to bottom!

Confronted by this kind of disturbance, by quakes in 'historical' duration, the sense of insecurity felt in our remotest
suburban outskirts is something each and every one of us
begins to feel.

And in the face of this amnesia of the moment, St
Augustine's existential question poses itself over and over
again: 'Is the real moment still present?'

If our answer is now negative and if the acceleration of
reality really has supplanted the acceleration of History so
dear to Daniel Halévy, then our learned doctor of the church
would find himself called into question!

Actually, if I'm no longer even aware of what time is,
incapable as I am, from the outset, of describing it explicitly,
Faith becomes a primary necessity, an urgent necessity, in
the face of the panic of total disorientation. In the face of
this ruin of time, all that then survives is the very particular
anxiety Kierkegaard talked about: 'Anguish is the possibility
of freedom. Thanks to faith, such anguish has an absolute
educative value. For it corrodes all the things of the finite
world and strips them bare of illusion.'[2]

Has duration, all true duration, become a run-of-the-mill
illusion, due to the acceleration of 'realism'? A lack of duration or, more exactly, this duration of lack which doesn't
even allow us now to grasp what is there, or who is still there,
overrun as they are by the untimely nature of what crops up
ex abrupto, of the accident that now replaces all events?

If so, what we will see after the amnesia of peoples with no
history is the senile dementia of a humanity finally globalised.
In the end, all the questions posed by the instantaneous
information revolution will be posed by the revelation of a
growing disappointment, in which immediacy and ubiquity,
attributes of the DIVINE, will not turn into attributes of the
HUMAN, after all, but will apply instead to an atemporary

inhumanity, whose crimes will never be condemned by any International Court of Justice – even if some claim to be preparing proceedings for just such an event: a sort of Trial of Galileo, that would not this time involve any ASTROLOGY-cum-ASTRONOMY, but CHRONOLOGY; the conquest of outer space mostly having shifted the question of science as a moral issue in the direction of this (spatiotemporal) continuum that the prophets of universal expansion go on about today.

Birth of time for the astrophysicists of the BIG BANG; birth of the instant for the anthropologists of the BIG CRUNCH of the present moment, this 'eternal present' of a relativity that was to suddenly become of vital concern to advocates of the geopolitics of places as well as of the history of the bonds societies are made of.

'The moment is uninhabitable just like the future', wrote Octavio Paz . . . A crisis in place (*lieu*) as well as in the (social) bond (*lien*), this 'no-place' (*non-lieu*) is about to become the major issue of the settlement involved in a TRANSIT currently overtaking the domiciliary inertia of our origins. Only, with this curious overturning of common sense which makes the sedentary homebody of today at home everywhere, thanks to the 'teletechnologies' of both the portable and high-speed transport, whereas the nomad of bygone days is now at home nowhere. It's as though everything was completely reversed and that, for want of any durable staying-put, circulating itself had been made habitable – in the process making cities and towns unsafe and fundamentally uninhabitable, with emergency lanes cost-effectively replacing the blocks of bygone neighbourhoods and the parking lot, along with our old public squares . . .

In the face of such an *anachronistic* state of affairs, now imposed on politicians of every stripe, the problem is no longer so much to do with the STANDARDISATION of the products and behaviours of a bygone industrial age, as the SYNCHRONISATION of sensations that are likely to suddenly influence our decisions.

In London as in Tokyo, for example, advertising displays along the avenues are kept up to date in real time, on digital screens linked by the internet. What is being trialed here is perfectly clear: people no longer synchronise watches, but the show! Any show of force – commercial or political – has to happen here and there, at the same time, in a 'present' that has no depth of field. The same logic is also behind the terrorist attacks that have been following each other, here and there, already for the last five years . . .

In Madrid in 2004, for instance, the attack that helped bring the Left to power in Spain was limited in its scope as a disaster by the fact that the five suburban trains targeted failed to arrive together at the scheduled time under the canopy of Atocha railway station, and this delay was not able to be anticipated in the planned synchronisation of the explosions to be detonated by the terrorists' mobile phones.

So the hypermodernity of real time is indeed the combined effect of the acceleration of History and of a shrinking of geographical space, triggering an individualisation of the destinies of each and every one of us, as well as of the various destinations of action.

Encompassing economic action and political action every bit as much as tactical and purely strategic action, mass individualism occupies the place once occupied by a collectivism now outdone by the new capabilities for treating our mental states, one head at a time.

The whole world stage is turned upside down as a result, to the point where *representations* gradually lose their pertinence – whether aesthetic, political or ethical . . . What is promoted instead is *presentation*, an untimely, out-of-place presentation that suppresses the depth of time for shared reflexion every bit as much as the depth of field of action and its displacements.

From that moment on, unlike in the theatre, where each spectator can see the unfolding of action that differs from session to session (the actors' performances), the cinemagoer sees the same film, from the same angle the shots were taken from, with only, as ultimate 'freedom', the freedom of not arriving on time at the start of the screening – a situation analogous to that of the terrorist frustrated by the lack of punctuality of trains.

With TELE-AUDIOVISION and the mobile telephone, we now not only see the same thing at the same time, but we can interact or, more precisely, inter-react, thanks to this allegedly real TEMPO which, in terms of the communication of interactive information, is akin to the radioactivity of matter and its ravages – that is, like an emotional 'fusion', or meltdown, of interlocutors as well as the 'fission' of being carried away, of a reflex action.

'Time no longer escapes History here, History has killed it,' the anthropologist Marc Augé specifies in relation to Chernobyl. 'Today only a catastrophe is likely to produce effects comparable to the slow action of time. [. . .] Sudden unforeseen death, the past is now dated: overnight, the place has been declared a desert.'[3]

Whether the RADIOACTIVITY of a place's matter or the INTERACTIVITY of information, the same bomb explodes at every instant, producing this transmutation in the historic nature of anthropology as well as in the physics of the infinitely 'small', which is now suddenly adapted to the infinitely 'vast' of astrophysics . . .

'Everything is happening,' Augé continues, 'as though the future could no longer be imagined except as the memory of a disaster which we only have a foreboding of right now.'

So, after the factual history of the Moderns, the time or, more exactly, the lack of time of an accidental history has obviously arrived. In it the culture of immanence yields its primacy to a culture of the imminence of disaster. And this disaster is no longer the acceleration of history described by Daniel Halévy, but acceleration of the reality of the present instant. ANTHROPOLOGY and DROMOLOGY merge the way the history of musical theories and musicology once did.

In fact, if a ruin is what remains of what once was in the past, today what remains of what happens, more than anything else, it is *ex abrupto*, this 'large-scale accident' that now dominates the event everywhere you turn, so to speak, in the contemporary world. This is not some malaise 'in civilisation' anymore, but a malaise in the very actuality of cultural events![4]

And so, from the 'objective' representation of events, we suddenly go to the 'teleobjective' presentation of a world globally damaged by a rampaging MEGALOSCOPY of ubiquity that not only distorts the history of civilisations and of art; it also, and equally, distorts the history of these ATEMPORARY sciences, erstwhile victims of the war of time, in which morphology now surrenders its customary importance to pure rhythmology for generic man – that biped who last century walked on the moon and now no longer walks anywhere except inside the image and its chimeras, those of an incontinent and desperate 'telesurveillance'.

Inhabiting the inhabitual every bit as much as harmful and uninhabitable instantaneity, the delocalisation of our activities also taints the realm of knowledge essential for life, in particular, social life.

In fact, if the anthropology of the instant suddenly becomes premonitory, that's because it is in turn undergoing what is known these days as 'EXTERNALISATION' – OUTSOURCING. As a reversal of perspective, outsourcing turns reality inside out like a glove. Accordingly, the 'vanishing point' in the real space of the Quattrocento turns into the 'vanishing point' of the real instant in an immediacy that is never anything more than a sort of

STROBOSCOPIC illusion scrambling all true perception and all true knowledge.

Actually, if the instant is uninhabitable yet inhabited by 'electro-technical' impulses, then it contributes, along with the past and the future, to the outsourcing of all retrospective, all memory, and so, all anticipation, to the point where premonitory fiction gradually steals a march on the history novel.

Something both mythical and magical then takes hold of these all too human sciences, once so attached to dates and places as well as to the chronology of past events; and it delocalises them, deterritorialises them completely. Everything happens now as though an imminent 'polarisation' was affecting all our different branches of knowledge, as though an activity that was once scientific and that once took place within the university arena had now freed itself, extrapolated itself in the exoticism of a globalisation 'for tourists' which is already allied to the discovery of an unknown planet, not to say an exoplanet unfit for life; an Earth foreign to the Earth, to its history as well as to its geography and, in that, similar to the one our astronomers are desparately battling to discover in cosmic space.

Faced with this exotic phenomenon, which we might call EXTRA-POLARISATION, the anthropologist of presentiment strongly risks morphing into an anthropologist of profound resentment – resentment towards the BLACK UTOPIA of a disaster that would force all of humanity, along with all humanity's diverse 'faculties', to vacate the premises and go into exile God knows where, far from the here and now, in the otherworld of cybernetic illuminism. This late sun worship, though, will not celebrate light, as in the not so distant past, but the speed of light alone. That is the new absolute of a century which, we all get the feeling, will then turn its back on the 'Enlightenment', degenerating instead into a dark age produced by a postmodern obscurantism in which the search for dark matter and its hidden energy, undertaken by our astrophysicists, would above all signal this denial of the visible and manifest that privileges the unseen, the unexpected, even the unspeakable in these 'clouds of unknowing' which, however, our learned doctors once derided.

In the face of this accident in knowledge that literally turns all the sciences arse over tit, stemming as they do from the Earth and its history, the emergence, the dazzling growth of ECOLOGY looks very much like a self-fulfilling prophecy for kick-starting an imminent ESCHATOLOGY that will not only be secular and atheist, but also neurotic. And *that* is far more frightening, even if we set aside the example of the Nazi notion of *Lebensraum*.

In effect, disqualifying planet Earth, the planet of the living, for reasons of insalubrity due to sundry forms of pollution – of distances as well as of substances – and setting off on a postcolonial discovery of a substitute somewhere else, would mean exiling all human science, that is, all EARTH science, in the hope against hope of discovering a perfectly exotic science no one has an inkling of, and pursuing an exodus from a Promised Land that clearly turned out not to be at the centre of the universe even though it managed, all the same, to have an exclusive hold on life and the 'life sciences'.

We might also note that the biologists, for their part, remain more materialist than the physicists, to say nothing of the astrophysicists. Here's one biologist who's not yet boasting of being an EXOBIOLOGIST: 'In this age of chaos and chance, in this age of fragmentary knowledge, we can no longer look at theories the way they did in the nineteenth century. The proof is that these days you get scientific theories and currents of thought, but no hard kernels of the sort Marx, Freud and Darwin were able to come up with. We're in the age of exploded theories.'[5]

You'll recall the humorist who once said: 'I leave my body to science fiction!' Today, if we're to believe the *savants* of EXOSCIENCE, it is the whole of the celestial body and its history that we are gearing up to leave to the heroic fantasy of the exobiologists, those promoters of the search for software that can do everything or, rather, redo everything. For the cloning of 'Mother Earth' by her sons can only ever be the fatal outcome of the cloning of her daughters' children.

II

The Waiting Room

Chapter 1

But when that which is perfect is come, then that which is in part shall be done away.

St Paul (I Corinthians, 13: 10)

Partial our science, partial our consciousness of its benefits – the globalisation under way causes the optical illusions of progress to disappear, one by one, even though that progress illuminated past centuries. Whence this principle of uncertainty before the risk of an accident in knowledge that will, in the near future, only add to the accident in the substances of an ecology that takes seriously Aristotle's second axiom: 'Completion is a limit.'

When the Accident becomes integral, when catastrophe turns into a system (an ecosystem) and disaster goes from being continuous to being a continuum – in other words, the space of time of a materialist reversal – what actually remains of a philosophical approach that unduly privileged the 'operative thinking' already condemned by Maurice Merleau-Ponty when he observed that, 'Science manipulates things but gives up the idea of inhabiting them.'?[1]

For 'science', as for its technologies, completion is therefore also a limit, an *ecosystemic* limit that removes the mathematical idealities of a knowledge where the 'constructive' pursuit of information passes itself off as being autonomous, with the quantitative numerology of information technology gradually coming to dominate the qualitative analogy of our vital perceptions. Whence the uncertainty of reason in the face of this accident in prediction which has now turned into prevention, if not simple precaution, but which will never constitute a sufficient reason – only an uncertain approach based on an estimate. And this approach will, in the near or not too distant future, create the need for a METEOROLOGY that is no longer exclusively concerned with the elements, the weather – temperatures, precipitations, winds, and so on; that is, with a patient search for 'models' of movements in the atmosphere. The new meteorology will need to try and come to grips with the mysteries, the enigmas, of the space–time of an accelerating dromosphere that escapes our projections, with its terrorising high pressure systems, its unknown economic storms and precipitations, to say nothing of climate change and its repercussions on the settlement of a terrestrial globe that has turned into a 'reduced-scale model' . . .

In this connection, we might for instance note that twenty years on from Chernobyl, the Ukraine has just endowed itself with a 'Ministry for Emergency Situations'. It seems it's not a matter of opening political ministries anymore there. Instead the State of Emergency sets up its administration in a bid to adapt to the 'fluid dynamics' that escapes the political economy of the region, just as radioactivity continues to escape from Chernobyl's historic sarcophagus.

Similarly, in the realm of insurance banks, big companies hoping to protect themselves against the economic ravages of natural catastrophes now call on geographers and climatologists. Essentially, though, these organisations prefer to rely on actuaries and engineers who have recourse to catastrophe simulation software.

And so, in the hope against hope of contributing to the prevention of the disasters that threaten said organisations, what we are seeing is the emergence of a new profession: the economic-disaster-modelling geek. A vision of the future, if ever there was one . . . No mass recruitment is anticipated for this type of job, though, for, as one expert points out, 'to make models, you don't need a lot of people'.

Among such specialists in negativity, some will devote themselves to natural catastrophes, others to industrial catastrophes or terrorist attacks . . . We should remember, though, that officially there are only three or four manufacturers of CATASTROPHE SIMULATION SOFTWARE in the world and that the advantage of such a situation for said experts lies in being able to use a sort of universal language, with the same software programme, thereby helping wrap up LOGICISM good and proper. That doctrine, developed by Gottlob Frege and Bertrand Russell, consists in bringing the logic of reasoning to bear on the psychological or sociological aspects of the phenomena under study.

Mythomania of a quantification encouraged by the constant development of information technology and of its drag effects on the requirements of a communication system in which the rapidity of the result takes precedence over its quality; a normative approach that seems to contribute to some 'advancement of knowledge' when, in fact, it drags us down into this digital dogmatism that poses considerable risks for democracy. Today the religion of numbers not only fuels opinion polls but election results, as we've seen with the ever-diminishing gaps between the candidates in recent elections in Mexico as well as in Italy, the United States and Germany.

LOGICISM is clearly poised to become the heavy industry of the digital age, as Maurice Merleau-Ponty already foresaw:

Operative thinking becomes a sort of absolute artificialism as we see in the ideology of cybernetics, whereby human creations are derived from a natural process of information, itself based on the 'human machines' model. There is today, not in science, but in a fairly widespread philosophy of the sciences, this entirely new feature which is that constructive practice takes itself to be and passes itself off as autonomous, and thinking is deliberately reduced to the set of sensoring techniques it invents.[2]

The phenomenologist concludes his warning on techno-science with these words: 'To think is to operate under the sole reserve of a controlled experiment where the only things that intervene are the phenomena our devices don't so much produce as record.'

To illustrate this observation, we might note a recent project whereby detection of major risks is reversed, since the computer in question is involved in producing said major risks. At the end of 2006, IBM effectively decided to build the most powerful supercalculator in the world, for the Pentagon. Installed at Los Alamos, the birthplace of the atomic bomb, this 'information bomb' will serve to simulate explosion of the nuclear weapons of the future.

To do so, it will use processors capable of up to *one million billion operations per second*, accelerating by as much the reality of the disastrous progress in weapons of mass destruction . . . Which prompts a personal question: after having resorted to meteorologists and other climatologists to calculate the economic risks of catastrophes, will the insurance and reinsurance companies one day have to call on the army and their new strategists to detect the major ecological risk of nuclear proliferation, thereby ushering in an emergency polemological meteorology whereby the political war between nation-states will give way to a 'transpolitical' state of emergency caused by an accident in history that will lead to chaos?

Fifty years ago in the newspaper *Le Monde*, the economist, François Perroux, delivered this judgement: '*Academic life on the whole is fairly impermeable to tragic issues. Geography, for its part, generally abstains in the face of major events, and it is especially for that reason that the discipline can't be considered a political science.*'

In fact, over recent times, the deep space race conducted by the military-industrial complex has enabled the development of an astrophysics which was to contribute to erasing the limits of the geophysics of the globe we inhabit.

In the twentieth century the issue of the completion of world conquest was not actually analysed in its ultimate consequences, in spite of the emergence of 'ecological' sciences and the development of meteorology. Whence the Babelian confusion over the 'nature of space', the space of the 'geophysical' atmosphere which is in no way similar to the space of the 'astrophysical' dromosphere, since, contrary to the other planets in the solar system, this one happens to harbour life, the molecular life of a space of time that has favoured History.

So, having omitted to consider outer limits, or completion for what it is, science then gives up the idea of inhabiting the expanse that saw it come into existence as material, as a field, and projects itself into the incommensurable immensity of an open system. There, astrophysics can look like some sort of progress in astronomic knowledge, but surely not like a solution to the crucial problems now confronting humanity. You would think the serenity needed for science could only be found now in contemplation of spectacles devoid of drama, devoid of historic tragedies, except for great cosmic catastrophes like the explosion of a 'supernova', or perhaps traces of the remote BIG BANG.

With contemplation via astronomic telescopes, followed by computer screens, having replaced perception of events, in anticipation of the arrival of domestic television, the TELEOBJECTIVITY of telescopes only strongly intensified the distancing, the retreat of the realities of that 'immediate materiality' that was, nonetheless, the very basis of our practical knowledge.

This is where the true utopia finally lies, in this omission of the territorial body that has seen the birth of all organic vitality. And this deterritorialisation heralded the deterritorialisation of the world of an 'actuality' that definitively eliminates, it would seem, the urgent need for a political geography of the civilisations to come.

'Let no one blame our isolation on destiny!' cried Hölderlin. 'We and we alone take pleasure in flinging ourselves into the night of the unknown. If we could, we'd abandon the realm of the sun and tear off beyond its limits.'[3]

That is now a done deal with the quest for these EXOPLANETS destined tomorrow to replace an outmoded planet whose obsolescence still saddens us, even though the idea of some sort of 'totalisation' is completely meaningless: 'Even if it were to go on for millions of years still, the world will end without being completed,' Merleau-Ponty warned us.

And so, science, all human science, is set shortly to turn into this state of continuous stupour that our phenomenologist of perception spoke about not so long ago in relation to the operative positivity which has ended today in the disappointment of the false imaginary of an 'electronic virtuality' that tries to fill the void, 'this regret at not being everything' – the patent failure of a universalism more scientologist than truly scientific.

It also explains the Catholic Church's recent soul searching over Faith and Reason, but indirectly also over reason and suicidal folly, this PHILOFOLLY of the worst-case scenario that worries those who subscribe to an age-old wisdom, confronted as they are by the chain of catastrophes now casting a shadow over history.

Two terms echo each other here: DETERRENCE and DISAPPOINTMENT. Deterrence, in the face of the scale of a disaster arising from the dazzling success of a science that has become subservient to total war and so to the probability of mutually assured destruction; disappointment, in the face of the major risks unconsciously produced by the prowess of our technologies of energy, information and, soon, genetics.

The Austrian biologist Edwin Chargaff writes in his memoirs:

> My life has been marked by two important scientific discoveries: the fission of the atom and the elucidation of the chemistry of heredity. In both cases, a NUCLEUS is molested: the nucleus of the atom and the nucleus of the cell. In both cases, I have the feeling science has crossed a line it should have stepped back from.[4]

But now a quite different NUCLEUS, it would seem, is being molested: the nucleus of a star that is, in its way, unique, precisely because it harbours 'biology' and its history. In the above observation, coming as it does from a scientist behind 'Progress', what is so very revealing is the connection made between the two terms, 'FISSION' and 'ELUCIDATION'. You'd think the latter, too, was merely the explosion or, more accurately, the implosion of a knowledge damaged beyond repair by its merits, its feats; the emblematic figure of an accident in scientific 'knowledge' that has just rounded off the accident in the 'substances' involved in nuclear energy.

Compulsory deterrence in the face of the threats of 'fission' as well as of atomic 'fusion', and outraged disappointment in the face of the risks of an 'elucidation' that would result, this time, in an INTEGRAL ACCIDENT triggered by the very positivity of these operative sciences that have allowed our modernity to flourish.

Actually, what is about to explode is indeed this other NUCLEUS, this 'bubble' of knowledge acquired over the course of centuries, ever since the business with Galileo; the bubble of a science which, like the globe, revolves not around the Sun, now, but around a blinding light 'brighter than a thousand suns': the light of the possible extermination of all critical awareness. This is a suicidal state for a mad science that has now reached the stage where the ECOLOGY

OF REASON and its philosophy are giving way to this
ESCHATOLOGY of a philofolly not unlike that heralded
last century at Auschwitz and Hiroshima and Nagasaki.

At the very start of the third millennium surely we can't
fail to see, at last, that the exemplary success of the sciences
of humanity, these sciences now so arrogant, fatally results
in the madness of an inhumanity that was not punished at
Nuremberg, but will be soon – by its catastrophic success, the
very scope of its 'extremist' feats. This will involve the return,
if not of the issue of a 'God, the Creator', then at least of the
possible finiteness of a 'humankind' that has taken itself for the
'Most High', the demiurge of a New World. And has done so
without even being conscious of the fact that it was making
uninhabitable not only the planet, but the whole of a science
more adept at manipulating beings and things, yet that, oddly,
gave up the idea of inhabiting this 'gigantic machine' André
Gorz recently spoke of, which, 'instead of liberating human
beings, restricts their autonomy, determining the ends they
are to pursue and how they are to pursue them.'[5]

Actually, the integral accident produced by this knowledge machine, which makes acquired knowledge unbearable
and the substances making up the molecular life of human
space uninhabitable, today results in a 'revelation' even more
momentous in its consequences than the whole raft of 'revolutions' that have delivered a modernity more transgressive
than really progressive.

Globalisation has outpaced Universalism at the very moment
when that astrophysics has made us neglect all true geopolitics
and all military geostrategy, along with the geophysics of the
globe.[6] Like the extravagant liberalism of a market gone single,
globalisation has denied the different cultures and their civilisations, in a unilateralism that even taints our view of the world.
This is where the politically correct bows out, overtaken as it is
now by an optically correct apperception that has falsified our

relationship to the real, as well as to the imaginary – just as Maurice Merleau-Ponty said it did. We've got to the stage now where the famous GLOBALISATION – *MONDIALISATION* – has turned into a simple MODELLING of our behaviours, in anticipation of the globalisation of affects – this synchronisation of our emotions that kicked off with the moonlanding and spiralled off considerably as the twenty-first century dawned, with the attack on New York, seen instantaneously by millions of individuals.

Let's hear what Derrick de Kerkhove, one of Marshall McLuhan's former assistants, had to say at a conference in Bordeaux in 2006 about such mindless modelling of our behaviours:

> The geographic expansion of our thinking largely spills over our national boundaries. This happens in an unconscious way, even if I believe *we need to overcome the notions of conscious and unconscious*. We're living in a world where we're endlessly confronted with globalisation. We can no longer escape a global dimension to being.

The loss of the sense of reality under way here is indeed a forced overcoming of the notions of conscious and unconscious, this shutting away – the 'great lockdown', as Michel Foucault called it – that turns the whole world into a lock-up universe, while the universe of the deep space of the cosmos appears like a horizon of liberation from the ecological constraints of the human habitat.

'I can well see,' de Kerkhove continues, 'that consciousness and the unconscious sort of merge in a way Freud didn't foresee. Google, the integral search engine, is the technical consciousness of this unconscious other that constitutes the digital person. *The technical consciousness is what you put on the screen*.'

As for the unconscious, that is the mass of the information circulating within the world of real space at the speed of so-called 'real time' and concerning you directly – or, we should say, LIVE.

So, it's pretty clear: the citizen of the finite world (the earthling) is now no more than a social reject, one of those outcasts OUTSOURCED by the sudden polarisation of terrestrial space-time, less generic than exotic, and for whom we need to come up with a new ethics, an ethics adaptable to the virtual space of the derealisation under way. It started with the invention of this so-called Sixth Continent involving perception that is itself digital and instantaneous, or as good as, with the screen-world of the Google Earth search engine completely replacing the horizon of phenomenological appearances.

With the global optics of teleobjectivity from now on replacing objectivity, *de visu* and *in situ*, the ocular correction provided by the lenses in our glasses (astronomic or other) is bolstered by the societal correction of training humanity's single gaze upon its fate, a feat brought off by perfect synchronisation of humanity's visual and auditory sensations. Once individual and personalised, such sensations are now collectivised by the tyranny of a so-called real instant that actually has strictly nothing to do with the *present* moment of our immediate consciousness, *hic et nunc*.

In short, we are dealing with a phenomenon of photosynthesis whereby the animal kingdom of our humanity is, all of a sudden, linked to the plant kingdom, in the cold light of waves that convey our images and our impressions . . .

In fact, given the photosensitive inertia involved in receiving messages, the INTERACTIVE televiewer, turned into an internaut in a space that stands in for the real world of avowed facts, tends to react in the same way as vegetal systems, the photonic illumination of screens more and more mimicking the effects of sunlight on vegetative life.

Whence the recent research into neuromarketing in this age of the televisiophony of the mobile and of the implantation of radio-frequency identification chips (RFID) within urban space, to the sole end of decoding the decision-making process involved in 'impulse' buying; truly subliminal practices that aim to trigger at a distance a purchase, a sale and, who knows, maybe even a vote . . .

Similarly, scientists at France's Centre national de recherche scientifique (CNRS) are currently working on image processing software capable of identifying every person who stops in front of an interactive billboard.

Going the way of the phenomena involved in interactive reception and their 'targeted' messages, will we soon turn PHOTOSENSITIVE, like the rolls of film that have disappeared from our cameras? Will we become sensitive to passing rays the way plants, flowers and fruit are to photosynthesis? Let's not forget, whatever happens, that the digital image used to be known as a synthetic image.

For proof of this new 'state of play', note that the nightclubs of Rotterdam and Barcelona are now applying the same system to their VIPs as was applied to Kevin Warwick, the first man to have received a radio frequency identification chip as an implant . . . A member of the British cybernetic service, Professor Warwick received his new interactive chip, which was connected to his nervous system, in 2002, at the moment that Sir John Eccles, the Australian Nobel Prize Winner for Medicine and Physiology, claimed that 'the brain looks to be more of a receiver of consciousness than a transmitter'.[7]

This explains the cosmic outlook of a 'global brain' open to receiving interactive mail in the age of instant telecommunications. The brain is a receptor, then; it has stopped being a dispenser of witty exchanges or of any consciousness of otherness or sharing of knowledge once meted out by the University, '*that "University of Enlightenment" now cut off from nature*', as Jean-Marie Pelt puts it, '*in which it has basked ever since the Middle Ages, knowledge of animals and plants having been banished from our university faculties, as alchemy and astrology once were under Colbert'.*[8] This has happened in spite of the 'realism' of a cyberworld so close to botanical photosynthesis and to zoological 'animality' and so far away now, accordingly, from the dynamics of perception that Merleau-Ponty observed as a phenomenologist, following in Husserl's footsteps.

In fact the greatest threat to science, to the exact sciences, is not so much the TECHNOPHOBIA of its critics now as the TECHNOPHILIA of its accredited promoters. Here's one – Claude Allègre, the French geophysicist and socialist: 'Mathematics is bound to go into an ineluctable decline, since we have machines to do sums today.'

Such declarations seem to ignore what comes next: now that software has eliminated mathematics and children all own calculators, from primary school on, numbers will no longer make any sense to them. LOGICISM will then take over from the ancient, the very ancient, religion of numbers!

So here it is, this Global Brain which, under the pretext of some digital progress shared by all, contributes to the launching not only of a normative 'metalanguage' now, but of a METASCIENCE that once and for all breaks with our acquired knowlededge – in the name of COGNITIVISM!

A perfect illustration of the growth of software is its spread throughout the domain of sporting competitions as well as within the domains of education and government administration.

As an exemplary prototype of a 'decision-making software support', the aptly named Zeus, can simulate in a few seconds all the possible courses of action in any given situation and offer the 'best choice' based on a multitude of offensive combinations.[9]

Destined for American football trainers, this software is equally interesting to supporters of Formula 1 racing or crews taking part in the Americas Cup . . .

Here's what one Americas Cup competitor has to say: 'The helmsman can receive up to twenty-five bits of information simultaneously. To avoid cognitive saturation, he needs to be supported in analysing and synthesising that data.'[10]

Comprehensive insurance is clearly not the only industry to use the 'synthesiser' anymore; prevention, until now simple precaution, is gearing up to dominate all human forecasting today – and tomorrow, no doubt, all decision-making based on free will!

At the Pew Research Centre in Washington they're even wondering about the future of the Internet network. In 2020, according to the experts, everything will be more and more visible to each and every one of us, thanks to the new system of surveillance and tracking; 'social overexposure' will reach the dizzy limit. Some are even predicting that before that date 'a radio frequency identification chip will already be implanted in neonates in the industrialised countries'. Initially intended to furnish medical data, these RFID tags will thus be used for the tracing and telesurveillance of individuals.

While traffic information was limited until now to spatio-temporal signals or road signs – from the tourist's geographic map to the video terminal – we have now got to the point where our displacements are guided by radio. Information technology no longer accompanies transport: it *is* transport. 'One day', says an executive at the Paris transport authority, 'we'll have shoes equipped with electronic maps' – a bit like the ones certain salesgirls have, with the heels recording the number of their displacements.

And so, after control of the OBJECT, it's over to control of the TRAJECTORY and the itineraries involved in the roaming of a living being who can still enjoy his or her own autonomy of movement – though not for much longer.

After the TELEOBJECTIVITY of our perceptions, our sensations and feelings, it's now the TELESUBJECTIVITY of our decisions that seems set to be conditioned. We might even compare this situation to HELIOTROPISM, which, in botany, determines the orientation of plants.

These questions all reveal how, today, interactivity is to information what radioactivity and its irradiation have always been to energy.

But if it's hallucinatory phenomena that we're after, we need to turn to the realm of cosmetics to get an idea of the future of 'messaging' in the age of photosynthesis.

In France, for example, researchers specialised in make-up are currently interested in the 'chromatic effects' of IRISATION, the way the colourless colour of soap bubbles looks different depending on the angle. In an 'impressionist' make-up, colour effectively derives from pigments: when these receive light, they absorb some of it and the light reflected and perceived by the eye then takes on the colour complementary to the one observed.

To manipulate such diffractions or interferences, this particular laboratory is resorting to so-called structural colour in which light has to effect a certain trajectory through minuscule patterns, all identical, which then produce surprising and intense colours at all points along the surface made-up.

The goal of the research is to endow the subject's eyelids and eyelashes, nails and lips with veritable (versatile) holograms which, through the play of interferences between colliding light waves, will cause a 'relief' to suddenly appear there – flowers, butterfly wings . . . unless, of course, the top model, adept at such PHOTONIC tattooing, agrees to sport the manufacturer's logo on her lips or her eyelids . . . .

Faced with such a profound identity crisis, the subjects of this photosensitive age now want to glitter for all they're worth, to become electro-luminescent. They are no longer content to be both 'impressionists' and profoundly impressed by the shock of the images and miseries of the times; they want to become luminous, dazzling. Whence the hunger for instant recognition, in real time. And with it, the HELIOTROPIC orientation of bodies moving before the cameras, this desire both to record and to be recorded without letup, thanks to these 'televisiophonic' mobiles – in the hope, not of being famous anymore, but of making famous, in a form of celebrity that has nothing to do with talent or lasting merit or with any kind of immortality.

With the new make-up, the old 'pictorialist' colouring of faces and bodies promptly turns spectral, acquiring a ghost-like quality, and the image of the person gleams in the fugitive manner of a plant, which flowers in anticipation of one day bearing fruit.

Besides, we won't understand anything about the return of the tragic, or the gothic craze, unless we see the photosynthesis involved in an emotional interactivity in which you are now required to appear in the distance and turn yourself into a sort of *iconic emanation*, the very word 'star' no longer being exaggerated but utterly trivialised, since it doesn't so much refer to a talented actor as to an IMPACT MAKER, a whole host of whom now clutter the postmodern stage of 'telereality', in the realm of art as much as in those of politics or terrorism.

To try to better grasp the astonishing bankruptcy of representation in the era of the photosensitive presentation of events and their organisers, let's take a closer look at that word 'MORPHING'. This Anglo-American term currently designates the continuous, animated transformation of one image into another; it thereby participates in kinematics, the 'energy of the visible' that is so important in these days of acceleration of reality where the audiovisuality of world views prevails.

Whether it is acknowledged or they try to fool us by masking it, what is at play here is a metamorphosis of age-old representation, ethical as well as aesthetic; perhaps even an endless anamorphosis of the tangible and the pertinent, in which acceleration of reality tends to prolong indefinitely, it would seem, the acceleration of the history of people's thoughts, as the disabused reflection of one Japanese man indicates: 'Omnipresent networks will also create omnipresent problems!'

'The hours will pass, but science will continue to grow.' This progressivist motto, written in Latin on the sundial of Cuvier House in the Jardin des plantes in Paris, illustrates Theodore Monod's notion that 'science is a lesson in Time' –Time, which either fades away or which we sum up thanks to History.

Today, as Winston Churchill judiciously pointed out at the end of the 1930s: '*We are entering a period of consequences.*' These consequences not only determine the politics of nations, but even more so, science: the human sciences as well as the whole panoply of our knowledge. So much so that, if science was once a lesson in time – in passing time as well as in the growth of 'Progress' – it is now turning into the science of the weather, day and night, a 'lesson in darkness'. Whence the scientifically uncertain nature of a climate METEOROLOGY that now has the world biting its nails.

If nothing is static in the 'dromospheric' world of the acceleration of history, logicism as a science, with its software that can do anything, today appears globally unreal. This has given rise to a series of controversies that are shaping up to be not only theological, like the opposition between Islam and Christianity, but also meteorological, like the opposition between those who claim global warming is catastrophic for the climate and climate sceptics, such as Claude Allègre, who not long ago thrashed 'the proponents of an ecology of helpless protesting which has become a very lucrative business for some people'. It has also given rise to the inevitable reconciliation between this dim view of climatologists and the fact that insurance banks are calling on them to forecast the economic or ecological disasters of the future . . . Still in the words of Claude Allègre, the former Minister for Research, 'climatologists don't have a clue what's going to happen because the climate is unpredictable since there is no forecast model that works'. A fine tribute to common sense, that; unfortunately, though, it is only useful as reassurance, not in tackling the risks of climacteric finiteness.

Curiously, according to the ex-minister and geophysicist emeritus, the issue for the new science is not so much tied up with knowledge, with uncovering, as with power and development; in other words, with a sort of will to power the size of the vital space of the life-giving star.

Such a will to power is, in any case, perfectly well illustrated by GEO-ENGINEERING and certain processes designed to counteract global warming artificially. These involve extremist practices that aim to innovate, in the near or not too distant future, a UNIVERSAL AIR CONDITIONING SYSTEM able to cool down the planet by means of aerosol injection of small particles high up in the atmosphere where, it is hoped, they will reflect some of the sun's rays; or by means of launching beyond lunar orbit an immense mirror that would sit between the Sun and the Earth, thereby adding a great artificial sunspot that might just be able to dim our planet's lighting . . .

As one climatologist, cautious about such practices, points out, 'it's so hard to evaluate the consequences of such large-scale manipulation, that some of our colleagues are pessimistic about the effectiveness of such measures'.

Having entered 'a period of consequences' in the twentieth century, *geophysical* science as well as the geopolitical economics of nations have suddenly been confronted with the limits of a completion: the completion of the star that bears history. Whence the megalomaniacal temptation of an ECOLOGY OF POWER and the denial, characterised by 'humility', of the humus at the origins of knowledge; this will to take out a kind of 'comprehensive insurance', not for some habitation anymore, for some kind of domestic accommodation, but for the very habitat of the human race. Meaning, a geophysical environment reduced to its most simple expression by specialised software primarily concerned with the state of a sky that is a lot more METEO-LOGICAL than meteorological, since tomorrow, as some people are already predicting, 'the computer will have disappeared as a singular object to become our environment'.

With the omnipresent modelling of behaviours taking over from international economic integration, the precautionary principle won't leave room for passive defence anymore, as it did in days gone by; it will only leave room for the passive security of probabilist anticipation, thanks to the use of 'statistical clouds' that now carry all prevention, thereby proving right the child who cried out before the fog: 'A cloud has fallen to the ground!'

Worried about the future of his discipline in spite of everything, the director of Météo-France said, just before the launch of the satellite, Metop, in the summer of 2006: 'If there were no unforeseen changes in the weather, if in the future the chaos barrier was torn down, the philosophical consequences would be absolutely disastrous.'

Mark Twain once said of the adventurers of the past: 'They did it because they didn't know they couldn't.'

Whatever our scientist-adventurers, those men who made the bomb and then went all the way to the Moon, might say today, now they *do* know they can't – and that's exactly what the accident in knowledge is!

Chapter 2

Unhappy the man who were to learn the secrets of his future!
He would feel impoverished before falling into poverty.

Seneca

Could knowing be opposed to thinking? 'How intolerable life
would be if we knew in advance what evils lay in store for us
in the future,' observed Church doctor and saint, Alphonse
de Liguori. In fact, if in the past contemplation of the world
and getting about in it didn't manage to make us believe in
the reality of its creation, and so, ultimately to believe in God,
now we can't even manage to believe what we know pertine-
nently, what History has nevertheless so harshly taught us.

This is doubtless the ultimate 'atheism', a sort of MONO–
ATHEISM; and this brand isn't even a more or less fruitful
'relativism' now, but a form of denial, of contestation of
acquired knowledge, whereby capital-D *Doubt* gives way to
capital-D *Deterrence*. This is not the aggression involved in
yesterday's military Deterrence anymore, either, but a civil
Deterrence, deterring us from access to a store of knowledge
accumulated over the course of the ages.

In this modern version of 'conscientious objection', people
are no longer happy simply 'not to believe their eyes'. It is
now a matter of denying what you know in order not to
believe anything at all anymore! This superior form of cretin-
ism passes itself off as some sort of avant-garde nihilism.

To illustrate this paradoxical situation, which is revealed
by the current crisis in the university, we might quote Adrien

Barrot, the higher education reformer: 'Teachers are the priests in a cult that no one believes in any more: the cult of truth.'[1]

And so, after the Accident in substances and the serial catastrophes that we look upon, powerless, the time of the Accident in true knowledge is upon us. This is a form of secular unbelief that is leading today to the sudden virtualisation of the tangible and the palpable and the pertinent.

The result is the current denial of the Judeo-Christian, and shortly no doubt the Greco-Latin, origins of Europe – this history of successive civilisations. All that is now negated the better to DETER THE FUTURE FROM HAPPENING, with its cataclysms that are all too predictable, in the realms of economics and politics, at least. In the near future, this capital-D Deterrence could even go as far as eliminating the D in Democracy.

With this period of convulsive inertia ushered in by the third millennium, the imbalance of terrorism actually serves to conceal an even bigger threat, a threat that can only be called progressive obscurantism. The 'negationism' peddled by the killers of the memory of the concentration camps was never anything more than an early warning symptom of this obscurantism, so powerfully denounced by the likes of Pierre Vidal-Naquet and Jean-Pierre Vernant, two historians who – and it's no coincidence – bore witness to the Greco-Latin origins of our different branches of knowledge.

In the end this reversal in a notion applied right up to the 'medieval' period in the history of our continent, now affects the whole body of knowledge taught at university and not just the philosophical and religious knowledge coming under Christianity anymore. This situation actually constitutes a sort of apocalyptic REVELATION of a crisis in the European university that is nothing more than a reflection of the loss of the values deemed UNIVERSAL in these times of instantaneous globalisation of affects. We're talking cathodic globalisation more than 'catholic' globalisation here – a phenomenon whereby the new orthodoxy of the 'single brain' is opportunely disguised behind the persistent notion of REVOLUTION: meaning, a telecommunications revolution that 'research promoters' never stop boasting is inevitable.

So the promotion of products is gearing up to take over, once and for all, from reason and the value of what is produced, to the sole benefit of this VIRTUALITY where the acceleration of reality has long outstripped that of History and its veracity.

With the reality effect dominating tangible reality on all sides, the objective memory involved in learning and the knowledge thereby acquired tends to fade away and disappear in turn, along with the truth of the avowed facts. What is promoted instead is this aesthetics of disappearance, the relativist denial that can only be called DETERRENCE . . . an allegedly 'civilising' Deterrence, today so cunningly rounding off the 'exterminating' DETERRENCE of civil peace in the age of nuclear proliferation.

In the face of such a threat, such a major risk of an integral accident in acquired knowledge, the necessity of preaching in the wilderness of amnesia makes itself felt. To bear witness to the truth, dissolved as it is in mass media logorrhoea, once again becomes an emergency – one no longer biblical, but clearly political – as a PRECURSOR to the tragedy to come. This will succeed the tragedy of salvation, in an age of a progressive obscurantism that uses all the tools of information, and especially of formation of the public, in its bid to deter the future from happening.

Whether it is the Hebrew desert, at the origins of the great monotheistic religions, or climatic desertification, the voice crying in the wilderness is space, is stone, is all matter!

Lacking any sense of the wilderness, any nostalgia for it, the West was fated to accumulate, to own, to develop, to build up goods, property, knowledge; the sciences, comfort, democracy were to make it clear that it was going to force itself on the world . . . with its museums!²

So wrote Henri Michaux while the Second World War was raging around him.

Now, the museum to come is a museum of an ecological and ethological, if not anthropological, disaster, with the unheard of possibility of a tragedy that will, this time, be genetic.

In imitation of John the Baptist, to bear witness to the truth becomes a categorical imperative in the face of the barbarity of a technoscience not only lacking a conscience, but which dissembles the truth and so profanes its origins in reasoning that is scientific, not 'techno-scientistic'.

After Marx's historical materialism, what we are witnessing, with globalisation, is not only the emergence of a progressive idealism of 'high turnover' information, but the even more perverse progressive idealism of public formation – education – in this digital age, thanks to denial of the knowledge once dispensed by the university.

For some little time now, in fact, we have been looking on, stunned, at a conditioning of understanding that aims to deter us from believing what we know pertinently: that nothing is ever acquired without loss of substance, without accident . . . and that there is no real knowledge without major risk.

Yet all we have to do is recollect a single phrase from the ascetic canon of those disciples of the Desert of the mind, who decreed in the sixteenth century that: 'No one gets lost without knowing it and no one remains in error without wanting to', in order to divine that the famous 'civilising deterrence' propelling the information technology revolution aims above all to dispense us from such knowledge, so as to lock us, if possible, in a lack of any will.

The will to oppose the damage done by a technical progress that just keeps on growing, without limits and without brakes of any kind; that spurns all physical finiteness, as though the dromosphere of acceleration of common reality allowed us to completely forget the sphere of the habitat of the human race – the cosmic inspiration for this 'astrophysical' delirium being found today in the exoticism of an endlessly expanding universe.

But let's hear it once more from Henri Michaux:

> In the year 4000, they'll read: I was twenty-four years old. I was on Earth and I was desperately bored. I'd circled many things; I'd circled the Globe, too, I don't know how many times, on weekends. [. . .] No, space is no more immutable, no more unfathomable or untouchable than any of the other gods. Trust me, you'll see, a new war is gearing up.[3]

Over the course of those years marked by the neurasthenia of total war, Michaux went on to write:

> For a long while already they'd been aching for a new explosive. So *they found an abstract on intra-atomic demolition.* The explosion, triggered by a few atoms of uranium, can be propagated in a chain reaction, like a powder trail. *Everything turns into munition for this*: water, stone, continent, atmosphere, planet, ultimately! Splendid fate of the imbecile: he can cut the branch from under him. This once unthinkable suicide is something humanity can do, is going to be able to do.

The poet signs off on a prophetic note:

> *To blow up Creation.* Now there's an idea that would please mankind: our answer to Genesis. At last a truly diabolical idea, what does God think of that? [. . .] Are we shortly going to blast the angels? If they exist, they can expect to be run through soon with shot, with atomic fragments and toxic shocks! *Let's get ready to hear space scream!*

After the author of a poem so appropriately entitled 'La Ralentie', or 'Slowing Down', let's turn to Martin Rees, accredited astronomer of the Royal Court of England:

> Physicists aim to understand the particles that the world is made of, and the forces that govern those particles. They are eager to probe the most extreme energies, pressures and temperatures; for this purpose they build huge and elaborate machines: particle accelerators. [. . .] These atom-smashing experiments [conducted in particle accelerators] replicate, in microcosm, those that prevailed in the first microsecond after the 'big bang', when all the matter in the universe was squeezed into a so-called quark–gluon plasma.[4]

And the universe thereby went from darkness to light.

A little further along, Martin Rees concludes: 'Likewise some have speculated that the concentrated energy created when particles crash together could trigger a "phase transition" that would rip the fabric of space itself. The boundary of the new-style vacuum would spread like an expanding bubble.'

After the music of the spheres that accompanied the Creation of the world, the implosion of the dromosphere would drown, in silence, the molecular life of the space of living things. That was precisely the theme of my essay *L'Espace critique,* which appeared some twenty years ago and which lent its name to my collection at Galilée, a publishing house dedicated to another astronomer, Galileo. He was the man who said at his trial: 'And yet, it revolves.' Today we need to retort to the people conducting the progressivist Inquisition: 'And now, it explodes!'

And so, from sphere to the dromosphere of the expanding Universe, the negative horizon is constantly stretching – to the great dismay of those of us who reject the optical illusion of a 'real time' that denies all – or almost all – reality to the real space of events.

By way of illustration of this implosion and its sly dissimulation, we might like to read the tale of the most dangerous atomic event since Chernobyl:

On 25 July, 2006, the nuclear power station at Forsmark in Sweden avoided turning into a scene of tragedy by a hair's breadth. Following an ordinary electrical short-circuit that caused a blackout in one of the reactors, the numbers on the controls started to go beserk and the workers found themselves faced with a machine that was not only out of control but had become uncontrollable. The batteries in the plant's four emergency backup generators failed as a result of the overall breakdown in the sector and the core of Reactor 1 heated up considerably. *In such extreme conditions, a major accident is only minutes away.*[5]

It took the emergency crew twenty-three minutes to finally get two of the four purpose-built backup generators to operate manually. According to the Swedish plant operators at Forsmarks Kraftgrupp, the first phase of destruction of the reactor's core would have occurred seven minutes later, and fusion, in the following hour, would have produced a leak of radioactivity that would have spread all over Europe. Even Lars-Olov Högland, who was formerly chief of construction on Reactor 1, claimed: 'It was sheer luck that there wasn't a meltdown.'

Luck? More like a miracle! When a simple short-circuit can trigger a nuclear catastrophe, it's everyone duck for cover, in the famous fallback room that power station architects are obliged to build for control-post technicians, whereas nothing is provided for the terrified hordes . . . unless the atomic shelters of the Cold War are reactivated – in peacetime, this time!

After all, in discussing the 2005 accident, even the owners of the Forsmark nuclear power station, Kraftgrupp, maintained that 'a nuclear reactor is in reality just a giant kettle'.

The key word here is not the word for the implement belonging to a 'tea service', but rather 'REALITY'; a reality singularly accelerated in the middle of the 2006 heatwave, for the twentieth anniversary of Chernobyl.

And so, while they never cease to tell us about 'foiled terrorist attacks' in, say, London, for over a generation they have omitted, it would seem, to foil atomic accidents – just so they can prolong indefinitely the lying by omission about a fatal risk that those who preach in the wilderness about a way out of nuclear energy never cease condemning. Actually the term 'omission' is not right here. We should be talking instead about 'lying by deterrence'.

More than a quarter of a century ago, in the face of the proliferation not only of atomic weapons, but of these nuclear power stations that are likely to cause mass destruction, another treaty should have been signed – 'against the pro-liferation of nuclear power stations'; a form of TNT against the ALL-OUT DETERRENCE of a scientifically revealed truth, this lying about the KNOWN that so clearly explains the disaster hitting the University, in the age of DE-TER, of all that is INTERRED, silenced. For not hearing what is being said by the ecologists is understandable, but no longer knowing what is known is something else, something whereby 'postmodern' negationism no longer only inters Auschwitz, but also Hiroshima and Nagasaki, to say nothing of Harrisburg or September 11, 2001 and the probable desti-nation of that famous Flight 93 . . .

'The whole land is made desolate, because no man layeth it to heart,' the prophet Jeremiah observed, a long time ago (Jeremiah 12: 11) . . . This is it, the catastrophe of so-called progress whereby an accident in science turns into an accident in all knowledge. Such an accident was a core principle behind the university, with the risk, there too, of a meltdown – core fusion – and the sudden destruction not only of 'higher education', but of all the knowledge acquired several millennia ago already, thanks to the various 'faculties' of the so often vilified Middle Ages. The current deterrence of knowledge, on the other hand, ends in this university of what is missing, where omission of the real yields to the realistic in a substitute continent, a so-called 'sixth continent'. This one comes with the lures and avatars of a virtual world whose trans-appearance is always just the reality effect of morphing – the 'fade-in fade-out' of sudden appearances and disappearances, in which rhythm wins out over form but especially over foundation, regardless of any 'morphology' of knowledge. The term 'dissolve' is more apt for this than 'fade-in fade-out', since visual and acoustic sensations, far from complementing each other, in fact meld in a magma, in the indistinctness of an 'art without end', an art without head or tail, in which audiovisuality achieves the chaos produced by the derealisation of the art of seeing and knowing.

This fusion/confusion at the core of the tangible is analogous in every respect to the effects of a (digital) Babel (babble) in which the world dissolves, finally, in indifference and inertia.

But to get to the bottom of this cold panic in sensations as well as in common sense and the evaluations its allows, let's go back for a moment to that word, 'morphing'. Morphing refers to the continuous, animated transformation of one image into another; it thereby participates in kinematics, that energy of the visible which is so crucial, with perception now relayed by these multiple screens in which the aesthetic conspicuousness of the real space of forms bows out before the real time of a TRANSFORMISM of the facts, promoting instead an optical illusion that is now going global thanks to the internet and its search engines.

In fact, if the invention of the cinematograph was already a spectacular exploitation of the energy of the visible, the subsequent implementation of domestic television was to play a big part in the acceleration of reality – that transformism of the formal and limited representation film offers into pure and simple presentation, whereby the imposture of immediacy was swiftly to borrow the whole array of media artifices in order to dominate all ability to see and all ability to hear, all 'contemplative recognition'. With the supremacy of the screen over the written word, this has spelled disaster for a university that has been passed down to us ever since the Year 1000. For in this so-called 'postmodern' era, automatisms are set up everywhere you turn – just so we no longer have to think about it, they tell us, but, in reality, so we no longer have to think at all, no longer write, no longer read, no longer count, mesmerised as we are by the illumination of screens that now dispense the deterrence of a once 'civilised' learning that tends, now, to disappear, along with the ethics of the real that we also inherited from a nobler age. An age in which teachers were only ever priests in a cult no one believes in any more: the cult of the truth.

One example among others illustrates such deterrence: the fact that faculty members have recently given up asking students for their reading notes on course text books, for the very good reason that such notes are already written up and, more often than not, available on the internet – which saves students from having to read the works in question!

You'll notice that, here, it's not the teacher, the 'priest', who has got the chop, but the – faithless – 'faithful', those followers who deliberately deprive themselves of reading and writing now that they are faced with the omnipresence of the screen, in a convulsive period in the history of the social sciences, in which the fixity of the internet takes over from the freedom of thought, and of movement, enjoyed by past students, those protagonists of a learning that involved real navigation between, say, Bologna and the Sorbonne . . .

The new brand of sensory deprivation has nothing to do with the 'solitary confinement cell' of torture chambers or with the 'sensory isolation tanks', once presented as the height of subliminal comfort as opposed to the noise of the city, the hustle and bustle of the common world, but that actually introduced us to the principle of a WAITING ROOM of a new kind, where everything comes to you without your having to leave, to move or to read, as you do with transport by rail. With this TERMINAL of a very high information flow of some hundreds of megabits a second, the energy of the visible, of the audiovisible, will take over in the near future from the energy of communal transport, thanks to fibre optics networks that will outdo the railways' hold.

But, by way of kinematics as well as of deterrence, the most conclusive example was provided on 7 September, 2006 by French President Jacques Chirac, during his official visit to the nuclear test simulation centre at Bruyères-le-Châtel in the Essonne region of France.

Paying tribute to the Commissariat à l'énergie atomique and to the Direction des applications militaires, which it administers, the president of the Republic declared: 'Our system of deterrence could never have established its credibility without these two bodies, which strive for the security and independence of France.'

In this regard, it is illuminating to note that, since signing the Test Ban Treaty (TICE) in 1996, France has launched itself into an ambitious simulation programme whose objective is to ensure the maintenance of a reliable and safe deterrence capability – in the long term, and whatever the cost.

So, the Tera 10 supercomputer, another 'information bomb', made by the company Bull S.A. and presented to Jacques Chirac in the summer of 2006, is just a device offering supercomputing fire designed to reproduce the phases of the explosion of an atomic bomb thanks to some 4,300 dual-core processors that each provide computing power of close to 13 billion operations a second.

Without equivalent in Europe and analogous to the most powerful American or Japanese machines, Tera 10 is clearly designed to best ensure the persistence of the deterrence of the strong by the weak and, especially in this period of uncertainty over North Korea and Iran, of the mad by the weak.

Fine proof, if proof were needed, of the 'acceleration of reality' that has just completed the acceleration of history in the Cold War, since it's no longer a matter of exploding the bomb here: it's all about its DETERRENCE; an all-out deterrence which, since 1996 and the Test Ban Treaty, now rests solely on the simulator kinematics of firing the absolute weapon.

More than a quarter of a century ago, when I wrote that 'speed is the world's old age', I did not expect to see such spectacular confirmation. In particular, I did not imagine that in 2006 the very name of the computer in question – this Tera produced by the aptly named Bull – would adopt the name TERRA, if by amputating a letter. The dromospheric bubble created by the acceleration of reality is now about to burst. Its explosion will illuminate a troubled period in History, in which the real time of globalisation will lock up the real space of the terrestrial expanse and throw away the key.

Since the industrial revolution in transport and transmission, in fact, KINEMATICS has been the source of our modernity – from the first filmed images cranked out by the cinematograph right up to this DROMOSCOPY provided by television and the whole digital-video toolbox. With this 'computer bomb', the energy of the visible is placed at the service of a terror-simulator designed not so much to ensure the defence of a national territory as the credibility of an ideal weapon, the *deus ex machina* of a secular society for which the notion of the 'credibility' of our nuclear strike capacity ought to be replaced by the term 'credulity': given that this is a period in History in which the notion of a 'great war' is suffering the same setback as a terrorism that no longer relies on the force of face-to-face armies, but on the chaos whipped up by warlords, in a private war masked by the standard of religion. This war makes abundant use of the spectacle of its panic-inducing effects, wherever the forces of metropolitan agglomeration have replaced the forces of geopolitical dispersal of bygone military campaigns and their age-old 'battlefields'. These are now resurrected in the perceptual fields of screens of all kinds – from those of the prophetic calculators of atomic test simulation centres right up to those of televised newscasts and of Google Earth, along with those of these mobile phones that can do anything, hear anything, see anything.

Lying by deterrence of reality, which resurrects lying by omission, is actually an attempt to institutionalise 'conscientious objection' or, better still, the duty of reserve – by misusing the energy of the visible, of the audiovisible, in order to condition people's thinking through the 'reality effects' of teleobjectivity. Yet such a retreat claims to be immediate connection. What they omit, here, is to specify something that nineteenth-century economists used to observe, with humour, regarding the ascendancy of the railway: 'The problem with the railway is that it works both ways.'

Distancing brings us closer. That's explicitly obvious, on condition, however, that we don't neglect what is implicitly obvious: the proximity of the far away greatly favours exteriority to the detriment of all conscious interiority. In other words, the 'real time' of telecommunications disqualifies the real space of objective presence, promoting instead the virtual telepresence of being over there. This has reached the point where the impact of the ascendancy of the railway of yore is now superseded by the impact strategy of the totalitarian enterprise of the multimedia over our mental outlooks.

Being here, here and now, is then subtlely discredited, rendered furtive, to the exclusive advantage of an absence imposed remotely by means of a teleobjectivity that no one contests in its optical credulity, or more exactly its 'electro-optical' (optronic) credulity. This is it, the sudden deterrence of present reality whereby we no longer manage to believe what we see with our own eyes, de visu; a derealisation of the world around us resulting essentially from the acceleration of a realist TEMPO. And that pace is greatly favoured by the 'light-speed' of waves that even exhaust geophysical expanse, in a pollution of the distances of the life-size world which has just rounded off the pollution of substances of every kind.

A phenomenological accident, this sudden loss of the reality of the real space of the geosphere leads us to neglect ecological objectivity, to the sole advantage of the 'real time' of this accelerating, allegedly economic DROMOSPHERE which neglects the nature of the material goods on which it was, however, based . . .

It's not some totalitarian UTOPIA that we're threatened with anymore now, as we were last century, but indeed an ATOPIA of a twilight of places that resurrects the twilight of the gods of Nietzsche's Valhalla.

What is now dying is not so much the divine, in fact, as the human, this living witness born of the humus, whose primitive humility has disappeared in the face of the extreme arrogance of these 'sciences without a conscience' which are ravaging not only matter, through its disintegration, but light and the speed of light. And doing so in a bid to harness the 'escape velocity' that enables us to escape from terrestrial gravity and to attain the 'exhaust velocity' that spells release from the environmental conditions of planet Earth. Which means denying this PLACE TO END ALL PLACES of all life, of all TIES, in the quest, an eccentric quest, for a surrogate EXOPLANET to replace the exhausted planet of History.

Note that it's not so much the air and the atmosphere we breathe that are polluted, but rather the geophysical space of this sphere that harbours us all. This has reached the point where the grey ecology of the contraction of time distances is soon likely to exhaust the vastness of the terrestrial globe before green ecology, in turn, notices the fatal pollution of the substances that compose it.

After the dead-end of elementary physics, whose seminal theories of general relativity and quantum mechanics remain irreconcilable, it is now revealing to see the sudden search for an initial error, the quest for a 'break in symmetry', which might perhaps produce the great theoretical unity so ardently longed for . . . To finally discover, thanks to experimentation, a break in space–time, a violation of Lorentz's symmetry – that is the basis of an international line of inquiry opened by one Allan Kostelecky, a British physicist by trade, working in the United States. The inquiry currently brings together dozens of research laboratories around the world in an effort to unmask the error, the ORIGINAL ACCIDENT, of physics. Such an error would go all the way back to Galileo, according to Kostelecky; in other words, to the very origins of a 'standard model' of space–time that Einstein merely inherited . . .[6]

More than ever in a critical situation, the space of our physics, like the space of our daily lives, is thus a mere construction; an architecture thrown up by perception of the phenomena that surround us and thereby shape, for all of us, this 'landscape of events' that makes philosophers and scientists mere humble landscape artists.

In this sense, Kostelecky is a bit like the Cézanne of the Mont Sainte-Victoire period, since, rather than grappling with the essence of things himself, the physicist proposes getting stuck into their appearance. Rather than attempting the great theoretical unification of physics by means of equations, he leaves it up to experimental phenomena to guide him towards a revitalisation of physics, fidelity to the small sensation à la Cézanne once more defining scientific experimentation. We might recall the Chilean cognitivist, Francisco Varela, who, before his premature death in Paris in 2001, made an urgent appeal for a return to the phenomenology of the sciences . . .

While we can't even manage anymore to believe what we see or what we know pertinently, deterrence of the self-preservation instinct is getting ready to invade the field of the basic sciences, having already invaded the field of vital experience.

Not so long ago, nuclear deterrence consisted in diverting an adversary from an aggressive intention, but the popular deterrence looming right now, with all these 'anti-terrorist measures', consists in diverting everyone's eyes from contemplation of a reality that still happens to surround us on all sides.

Because, as everyone of us knows, speed is not a phenomenon, but the relationship between phenomena (their relativity). 'Not to believe your eyes anymore' is no longer so much an exercise in 'conscientious objection', the doubt that is so fruitful in philosophical terms. From now on it is an exercise in objection to science, to all knowledge, acquired, as that has been, though, with such difficulty.

This explains the sudden untimely and misplaced emergence, in the West today, of the very latest ATHEISM – an atheism trumpeting Reason and the Universal – that now confronts MONOTHEISM and, in particular, Catholic Christianity. A totalitarian form of a fatal unbelief which we could describe as MONO-ATHEISM, this is an atheism for those who no longer believe in anything, in any part of this ALL of their immediate environment, in the face of the sudden acceleration of a surrogate reality that tries to suppress the primary importance of the phenomena involved in perception of the world. Its slogan could well be the one touted by internauts: 'LOOKING HAS NOTHING TO DO WITH IT.'

'Electric' energy or 'atomic' energy . . . Oddly, the industrial revolution's prophets of progress have overlooked, it would seem, the historic importance of 'kinematic' energy; this DROMOSCOPY, this energy of the visible and the audible that nevertheless conditions the revolution in physical transportation as well as in instantaneous transmission.

Implemented in the nineteenth century with photographic instantaneity, which allowed the cinematographic unreeling of the film, dromoscopy was extended, in the twentieth, with the videographic boom in analog television. The current digital computerisation of the process is merely the ultimate revelation of an energy of plausibility – wherever real space gives way to the 'reality effect' of rhythmic sequences, in a hypnotic present they nonetheless dare to call 'real time', even as the objectivity of our perception *de visu* is confiscated in favour of a sort of teleobjectivity in which events are endlessly telescoped.

Now that dromoscopy of the race of phenomena is winning hands down over scopy – optics *in situ* – we are looking on, powerless, or as good as, at the postindustrial staging of an audiovisible energy whereby rhythm wins out once and for all over both form and foundation. The latest generation mass telecommunications tools condition a reality which is, this time, virtual, whereby INTERACTIVITY is to 'general public' information what RADIOACTIVITY was to energy in the nuclear arena. The disintegration at issue here is that of the space–time of the phenomena involved in an instantaneous perception that conditions all experimental reason and all religious faith at once. The perceptual faith of this kinematics anticipates the effects of belonging that the old religions used to provide, and that is why it is so hypnotic. A veritable illuminism, stemming from the speed of electromagnetic waves, kinematic energy thus entails the supremacy of the rhythm of the morphing of sensations over the morphology of points of view. The importance given to the reality effect provided by the TRANSFORMISM of sequences ends in this postindustrial synchronisation of emotions that caps off the industrial standardisation of opinions and products. This synchronisation of emotions has the GLOBALISATION OF AFFECTS as its horizon, but that would just be the dizzy limit of the tyranny of real time; a modelling of behaviours that the current globalisation is merely trying on, anticipating tomorrow's collective HYPNOSIS of pure presentation, which will make up for any representation (aesthetic, ethical or otherwise). Our old freedoms of impression, just like our freedom of expression, will disappear in the face of the photosensitive inertia of any individual whatever. That individual will then be nothing more than developer for camera shots extending as far as the eye can see, in a DEREALISATION of the analog that is already foreshadowed by the considerable boom in the DIGITAL, whose advertising adage in the near future could well be: *DON'T* MOVE ALONG NOW, THERE'S EVERYTHING TO SEE BUT NOTHING TO UNDERSTAND!

III

Photosensitive Inertia

Chapter 3

Memory is made up of static shots.

Susan Sontag

With the primacy now accorded to light or, more precisely, to the speed of light perceived as a cosmological horizon, we have entered a new order of visibility in which the temporal perspective undergoes a mutation. The passing time of chronology and of history is suddenly replaced by a time exposed at the speed of electromagnetic waves.

This drift of scientific absolutism from Newtonian space and time to Einstein's absolutism involving the speed of light is in itself 'revealing' in the strictly photographic sense of the term. For the physicist, as for the theologian, 'time is the cycle of light'.[1] The temporal order, so dear to Kant, becomes the order of speed with the father of general relativity. According to Einstein, in fact, 'nothing in the Universe is fixed' – except, we might add perhaps, memory.

Actually, the order of absolute speed is accordingly an order of light, a sort of LUMINOCENTRISM, in which the three tenses – past, present, future – are reinterpreted in a perceptual system which is no longer exactly the same as that of chronology, but belongs rather to a CHRONOSCOPY. Here, with Einstein, an order of succession according to Leibniz, becomes an order of exposure, a system of representation of a Universe where future, present and past become the conjoined figures of under-exposure (the future), exposure (the present) and over-exposure (the past).

In any case, this issue of representation in physics caused a conflict between Niels Bohr and Albert Einstein very early in the piece. For Bohr, the notion of a trajectory of particles no longer makes any sense in quantum physics or, at least, is no longer useful, whereas Einstein maintains the importance of all sorts of vehicles (trains, lifts . . .), and we can see very clearly how serious the loss of the notion of a path or trajectory (and so of geometry) would be for him.

A distant heir to the 'ballistic' relativity of one Galileo, Einstein could not actually accept the visual conjuring trick of quantum mechanics. For Einstein, as for many others, speed serves to see – as a qualitative magnitude, a primitive unit of measure, preliminary to any other geometric carving up or chronometric reckoning. Speed is the light of light!

This is something that photographic experiments with different exposure times were to confirm, from the darkrooms of Niépce and Daguerre, through the chronophotography of the likes of Muybridge and Marey, right up to the most recent elementary particle accelerators, those telescopes of the infinitely small.

Kant's philosophical thesis, according to which time cannot be observed directly and is in the end invisible outside the aging process, collapses, since relativity, Einstein's 'theory of the point of view', now corresponds to a sort of 'photographic' – or more exactly 'phototonic' – focussing on the world.

Once, 'letting time pass' served to see time pass indirectly, duration only ever being a gradual unveiling of events. But, with Einstein's relativity as 'generalised optic exposure', it is not this 'gradual' quality – the extensiveness of time – that allows us to see, but rather the 'intensive' quality, the maximum intensiveness of the speed of the light of waves, that does the job.

From now on, the 'light of time' is no longer the light of day, the light from a more or less radiant star. It is the light of the speed of photons and their radiations – a light perceived as a standard, the ultimate cosmological limit of the universe.

Not so long ago, passing time still corresponded to an 'extensive' time, the time shown by ephemerides and calendars – which fully justified Kant's thesis about the invisibility of time. But today's instantly exposed time corresponds, on the contrary, to an 'intensive' time, a time belonging to the CHRONOSCOPY of the relativist eternal present; an integral optics associated with the ubiquity and simultaneity of a divine gaze, *totum simul*, or 'everything at the same time', where the successive moments of time are co-present in a single MEGALOSCOPIC perception that turns moments, once successive, into a veritable 'landscape of events'.

So the DAY of general relativity is no longer the day created by the cycle, the revolution, of the sun, but rather a day of phototonic resolution, an astrophysical revelation that will, finally, allow a generalised readability of different durations – the visibility of time – just the way ocular accommodation or the focussing of a high-resolution lens enhances the crispness of a shot, a snapshot . . .

This is the impetus behind today's race for high energies, the building of gigantic accelerators, like the collider ring twenty-seven kilometres in circumference, built by the Centre européen de recherche nucléaire (CERN), in Geneva; or the new proposal put forward by certain physicists, anxious about the way experiment lags behind theory, to build a particle accelerator right round the Earth and to build an even vaster one in satellite space – just to enhance the brightness of the light of speed even further. This would mean the dawn of a subliminal day, bearing no relationship to the race of the Sun from sunrise to sunset; the coming of a sort of durationless duration, of an 'intensive' time likely to supplant the extensive time of History, once and for all this time.

Following the atomic disintegration of the space of matter which, with nuclear proliferation, has landed us today in the

critical situation we are all familiar with, the phototonic disintegration of the time of light seems to be upon us thanks
to great colliders. And this disintegration will shortly entail a
considerable cultural and political mutation, whereby depth
of time will win out over the old depth of field inherited from
the perspective space of the Quattrocento. The term 'transparency' won't then so much cover the appearance of objects
made visible in the instant of looking, as appearances instantly
transmitted at a distance. Whence the proposed notion of the
TRANS-APPEARANCE of 'real time' intended to revive
the notion of the TRANSPARENCY of 'real space', with
the live transmission of images of things now compensating
for the former appearance of air and water, and even the glass
in the lenses of our telescopes or 'photographic' cameras.

It's not hard to see that the implosion of real time today
more or less completely conditions our exchanges and our
various activities, as well as our perception of the world.

This amounts to a veritable TELESCOPIC CRASH
and it heralds other interactive catastrophes; an economic
crash, certainly, but especially some disaster in social
communication, with repeat divorces largely affecting
INTERSUBJECTIVITY between people, between individuals, since the more the speed of information increases,
the more control tends to grow, the all-pervasiveness and all-
seeingness of such control aiming to turn our much-touted
CYBERNETICS into a substitute for the human environment, into our natural habitat, our 'New World'.

Note what Marc Augé has to say as an anthropologist of
the present:

> We have never been as close as we are today to the real, tech-
> nological, possibility of ubiquity. [. . .] This time, we will
> be able to handle immobility, but will we still be travellers?
> It's certainly not for nothing that the metaphor of travel is so
> often associated with cybernetic activity: you surf, you travel
> on the Internet.

Yet, Marc Augé goes on to say: 'Communication is the opposite of travel, since communication means instantaneity, whereas the traveller takes their time, hopes and remembers.'[2]

In this sense, we might even add that contemporary tourist practices have more to do with communication than with any long-haul voyage. And we might also refer back to the experiment of one artist-traveller of the early twentieth century, as an example. László Moholy-Nagy tried his hand at abstract painting – photography without a camera – to practice what he called light-painting, placing objects directly onto photosensitive paper and exposing them to light to create perfectly 'ghostly' compositions that he called photograms. The internaut, on the other hand, sitting there facing a screen trained on, say, Google Earth, practices a sort of on-the-spot trip, geographical transport without a vehicle – but not without an engine, thanks to the speed of a search engine that supports the photosensitive attention of the impenitent *voyeur-voyageur* of the Internet.

'Technology has declared war on daylight', noted Dietrich Bonhoeffer as the 1930s and Ernst Jünger's '*Totale Mobilmachung*', Total Mobilisation, got under way . . . But, in actual fact, it was the photographic lens that opened the aperture of hostilities back in the ninteenth century. And why, while we're at it, the term 'PHOTO-GRAPHY', which privileges the recording of the image to draw attention to the 'shot'? Why not 'PHOTOSCOPY' instead?

In its own way, though, the apparatus in question is exactly like the TELESCOPE as well as the MICROSCOPE. All three allow you to see, to contemplate, the depth of field of perceptual space. The PHOTOSCOPE, on the other hand, acts within the time depth of photonic radiation, the space–time of the perspective of the light of speed.

Ever since the revolution in (industrial) transport, inertia has had a very bad press. 'The inert man gets in his own way',

Seneca wrote, as you may recall. While Einsteinian relativity taught us that, in physics, you can't separate space from time or matter from energy within the 'four-dimensional' continuum, that doesn't stop the solid and its associated statics from being relentlessly devalued. What gets promoted instead is dynamics (of waves, of fluids), within the order of a now automatic and digital perceptual process whereby 'movement is all and the goal, pointless' (or as good as). The inertia of the object is discredited in favour of a trajectory endlessly accelerated – so much so that it has now reached the limit of the speed of light in a vacuum.

We might recall what Marcel Duchamp said about the urgent need for an extra-retinal art, an 'art as far as the eye can see', that would finally free itself from the retinal and photoscopic persistence of the human eye.

We should remember this: for the wheel to turn, it has to have a fixed point, a hub that doesn't turn. So, just as retinal persistence is the basis of our perception of movement, of the kinematics of the world's animation and, with it, the invention of the cinematographic, photosensitive resistance is the 'hub' of the 'photoscopy' in question. In other words, of pictorial appearance as well as of photographic revelation. This is something a colourist like the painter Seurat understood perfectly, unlike Degas and others, including Duchamp, to say nothing of the 'Futurists' . . .

In fact, if drawing is the virtue of painting according to Cézanne, the fixity of the photosensitive inertia of the shot taken by the 'recording chamber' is the virtue of photography every bit as much as cinematography.

Far from being a handicap here, something missing in relation to the sequence of animated images, the initial inertia of the photographic act is its future, its truth. And the 'art of the motor' (audiovisual) of the camera for automatically recording movement, an art that took over from the 'art of the actor' photographer, is nothing more than the past of an illusion; an optical illusion that dims – who knows? maybe forever – the light of Day.

And so, the apparent fixity of the snapshot which once seemed like a delay, an incompetence in relation to the unreeling of the sequence produced by the photogram, is much more a photostatic foreshadowing of the inertia of reality, of its memory, in the age of implementation of the illusory ubiquity of a 'real time' that claims now to supplant the real space of the geosphere of a world still 'habitable' – at least for a while.

'Immediacy is an imposture', Dietrich Bonhoeffer further warned . . . So is the instantaneity of our teleobjective perception, and we are very soon going to find that out to our cost. Whether the various propagandists promoting the motorisation of objective perception like it or not, 'contours' and 'contrasts' are crucial aspects of the photoscopic shot. To fix the limits of what appears and master contrasts in tonal values is part of heliogravure as well as *gravure*, engraving. That is the danger Nicéphore Niépce feared when he described the 'devaluing of solids whose contours get lost', becoming blurred in the 'artistic haze' that was about to prevail, with the industrial innovation of the standardisation of the film sequence.

When the gifted inventor of the act of photography fears the loss of value of the solid, he is basing himself, as the engraver does, on the importance of statics, of material resistance and the retinal persistence of the object's contours; to sum up, the importance of memory much more than the revelation of the image in the shot.

We can all easily verify this with our own eyes, anyway. Just stare at a light contrast, then shut your eyes; what pops up is not so much a detailed image as the outline of the object standing between your eyes and the light source.

Here again, the issue is not so much to do with the more or less coloured image as with its profile and the way the contour of the interposed solid stands out like a *découpage*, a cut-out, in the illumination that reveals it to our eyes.

Let's hear now what the painter, Henri Matisse, had to say about colouring: 'What counts most with colour is relationships. Thanks to relationships and to relationships alone, a drawing can be intensely coloured without colour needing to be put in.'

That says it all about the appearance of the graphic movement and its 'colour enhancement' – but, equally, about 'photographic' revelation. It would all be forgotten later, though, firstly with the spectacular innovation of a cinematograph that owed more to Daguerre than to Niépce, and then with the Lumière Brothers' 1903 invention of colour AUTOCHROMES.

In fact, the first *découpage*, the one that leads to the *montage* of perception *de visu*, is indeed the one made by the contours of solids and their different contrasts in tonal values, which make up the viewer's binocular image. Here again, the endlessly accelerated movement produced by the photogram and, later, its more or less extravagant colouring, were to definitively outpace the heliography of retinal persistence, that universal writing of light that allowed the birth of kinematic energy and the filmography it spawned.

Colour, tonal value . . . The principle of contrast is clearly well and truly behind the aesthetics of pictorial appearance, as well as photosensitive revelation; and when Niépce fears the devaluation of the contours of solids, he flags the extreme importance of *tone*, not *colour*, in the sudden emergence of PHOTOSCOPY. Whence the analogy with the line drawing, followed by engraving; but equally with the magic that emanates from the 'still life': a subject of predilection for painters and artists, especially from the Renaissance and right up to the nineteenth century, when the birth of photography would only prolong this fascination with the inanimate object.

How else can we interpret the use of the word 'OBJECTIVE' (LENS) to designate what was still only the

PINHOLE of the camera obscura, except, of course, in terms of this interest in the lens as a 'solid object' – but especially in the thing placed in front of it: *OBJECTUM?*

'Initial resistance' of solids and their contours, or 'retinal persistence' of the eye of the onlooker, photography has more to do with the still life, and so, with the OBJECTIVE object, than with the TRAJECTIVE, the trajectory of the unreeling film.

In this sense, any contrast in tonal values is the equivalent of the *'contrechamps'*, or reverse angle, as well as of the *contre-temps* of découpage and montage in film. Both are merely the result of the persistence of the object's forms. Niépce sensed that this persistence was indeed the source of the photosensitive revelation.

A contrast not essentially of colour, this 'concretion of visibility', as Maurice Merleau-Ponty called it, is firstly, though, a contrast in tonal values within a figuration of solids that takes us back to heliogravure and to its similarity with engraving on plate; copperplate being the essential 'support surface' that was soon to provide proof, *la preuve,* in the form of the photographic print, *l'épreuve photographique*, of the subject's retinal persistence – the basis both of the writing of light and of its translation into sequential movements.

So, thanks to the invention of Niépce and Daguerre, the form–background relationship of the configuration of solids was to prevail over the relationship of light and shadow that marked the show projected by the magic lantern.

Thanks to those two inventors, the contrast between full and empty will lord it over the immemorial play of shadows. Interposed in the interval separating the contours of solids, this perforated REVERSE FORM will act in the manner of a CLAUSTRA, or OPEN-WORK WINDOW, on perceptiveness, the subjectivity of an eye that will effectively turn into the 'window of the soul' of the subject, in the photoscopic glimpse of a passing illumination that will owe everything to the opacity of the inert object.

Disqualifying inertia, fixity, Seneca wrote that 'the inert man gets in his own way'. Extending this observation, the contemporaries of the motorisation and industrial standardisation of vision were soon to discredit, in turn, the inertia of the one-off object made by hand, promoting instead its serial reproduction. This paved the way for the myth of general mobilisation and the panic-driven acceleration of a movement in which the speed of light waves was to favour the pre-eminence of time, the real time of telecommunications, over the all too real space of perception *de visu* and *in situ*, only to wind up, finally, in the discrediting of all 'geophysics' of solids.

We then saw a new form of identification, in which aerial photography itself turned into GEO-GRAPHY in the literal sense of writing inscribed on the surface of the world; the inertia of the terrestrial globe's solid contour inscribing its geospherical reality beyond erstwhile cartography, in the inertia of an image that confirmed what André Disderi was talking about when he said, in 1862, that 'instantaneity is the spot of precision in the representation of nature'.

From that date on, the opposition between objectivity and subjectivity stopped being enough for us in our discrimination and, thanks to the labours of Muybridge and Marey, this led to the emergence of the famous trajectivity that was to usher in the boom in the trajectography of solids, in particular of cruise missiles . . .

Objective, subjective and, lastly, trajective, the perceptual prism then opened out into the fourth dimension, doubling the age-old perspective of real space (of the visual) with the perspective of real time (of the televisual), while our 'objective' perception is now clearly distinct from the 'tele-objective' perception of the world's different 'vision machines'; the spot of precision of instantaneity (tele-audiovisual) secretly completing the vanishing point of the space depicted by the Quattrocento perspectivists.

Curiously, this revolution in the relativist point of view, and with it, our view of the world, far from meaning progress in discrimination, has totally disqualified the primary importance of the fixed point, in favour of a vanishing ahead of all points (pixels). It thereby brings on a gigantic optical illusion that will soon affect the geopolitics of nations, inducing the five continents to give up their geophysical reality to the advantage of a Sixth Continent, this one virtual. The megalomania of the madmen of yore will bow out before the MEGALOSCOPY of an all-out ubiquity that will lead the different powers into error, if not, in the near future, into tragedy.

The Gallo-Roman poet, Namatianus, hailed Rome in these terms: 'You have turned what was once a world into a city.' A world city, as they like to claim these days, delighting us with such anamorphosis, even though it is now merely a matter of a teleobjective aberration stemming from an eccentric satellite viewpoint, analogous, all in all, to the gravitational mirages that split the image of celestial objects which the adaptive optics of our astronomical telescopes allow us to admire anyway!

In the end, the GLOBALISATION currently nearing completion is only a STROBOSCOPIC optical illusion, the modelling of a world whose contours and contrasting tonal values seem to have been eliminated to the sole benefit of a negationism that is now 'geo-graphic' and not just historic anymore. For some are claiming that 'the Earth is flat'[3] and that it would be good to flatten it even further by indefinitely accelerating its reality or, more precisely, the megaloscopic telereality not only of an idea but of a *single* undifferentiated *vision*.

Yet, while certain Europeans sit around deploring 'the immobilism besieging us' in contrast to supposedly more 'dynamic' nations, you only have to look closely at the historic fixity of the different political structures to sense, *a contrario*, that the 'futurism' of the *autostrada* that did not need the automobile to come along before organising the trajectography of the Roman roads scoring the Latin continent. A

very real continent ever since 'Christianity' began, Europe
is confronted today by the stroboscopic illusion of a virtual
continent spawned by the cybernetics of the Single Market,
with its repeat economic crashes, in anticipation of the great
torment, the great test, of an integral accident.

Strangely, though, while North America, protected from
the Second World War as it had been, wallowed in the serial
advertising of the likes of Warhol and company, ravaged
old Europe threw itself once more into kinetic art. The idea
of the 'motor-eye', Dziga Vertov's CINE EYE, took off
again – only, it was displaced from 'futurist' kinematics to
the statics of an artwork, of a plastic environment. European
artists contested the supremacy of film's aesthetics of disappearance over the aesthetics of the photosensitive appearance
of memory's static frame, which had prevailed from cave art
right up to the photographic plates of a heliogravure that had
once again found, in the darkroom, the perfect substitute for
the walls of prehistoric caves.

'Photos aren't art, they're just snapshots', observed Lord
Snowdon, who just happens to be one of the photographers
of the Royal Court of England . . . During that historic
period when the great cities were being rebuilt after having
been ruined, in an instant, by 'fire from the sky', the metropolitan agglomeration spread beyond its usual perimeter
boundaries, once again proving Nicéphore Niépce right in
dreading the devaluation of solids whose contours get lost.

TABULA RASA of the bombing of large urban areas,
rural, then postcolonial, exodus, the fixed point of urban
perenniality vanished bit by bit in the nebulosity of conur-
bations, those metastases of a trans-historic and soon trans-
geographic resettlement.

An anecdote illustrates this growing precariousness. It
is provided by the farcical proposal put forward by urban
authorities in the United States for the future name of the

gigantic conurbation that sprawls along the East Coast all the way between Boston and Washington. How, indeed, can you rationally baptise such an interspace?

The solution they came up with and later abandoned was BOSWASH, which would have turned the megapolitan nebula itself into a sort of sequence between its two inert extremities by splitting its proper name in two, through scissiparity!

Anyway, it's hard to see why such a pairing didn't start with the South, in other words, with Washington, the capital of these States, themselves United. This would have given us WASHBOST. Unless the name of the East Coast nebula were to change configuration depending on whether you approach it from the North (BOSWASH) or from the South (WASHBOST) . . . A question of semantics, no doubt, but why not of kinematics, given the now predominant influence of traffic flows?

Everything changes, as a matter of fact, depending on whether or not *seeing* can be coupled with accelerated *moving*, right up to the point of attaining the inertia of the speed of the light of waves. What was already true of nascent heliography is every bit as true of cadastral cartography and the kinematics of the highway in the era of general motorisation.

In fact, when the space interval between urban districts and the city centre yields its historiographic primacy to the time interval of an eccentric mobilisation, then the PERSISTENCE OF THE SITE, which is what founded the city in the first place, tends to disappear completely. This calls into question not only the 'city state', as in the not too distant past, but also the nation state and its representative democracy as part and parcel of a settlement that sought to be constant. What we get now, instead, is chaos, unparalleled tyranny: the tyranny of the 'real time' of an INTERACTIVITY that is every bit as forbidding, in relation to information, as nuclear RADIOACTVITY was in relation to energy.

We can now better grasp the seemingly paradoxical parallel between the invention of photographic instantaneity, which ushered in the age of cinema in the nineteenth century, and the decline – heralded from the very start of the century that followed – of 'urban persistence', thanks to the AUDIO-VISIBLE energy of the illuminism of telecommunications.

The devaluation of solids, the progressive erasure of their contours, will thus be succeeded by the disqualification of the old citizenship, '*droit de cité*', along with the famous 'Rights of Man and the Citizen'. Instead, what is pushed is a sort of artistic blur or, more precisely, a media blur, that ends in transpolitical chaos on a planetary scale, cunningly hidden behind this brand new will to power that goes by the name of 'globalisation'.

'Why do the heathen rage, and the people imagine a vain thing?' the Psalmist was already asking. 'Why this craving for nothing, this running after illusion?'

After having so long sung the praises of the folly of an acceleration without any limit other than that of light, we now urgently need to 'sing the praises of inertia'; of this statics, this photosensitive fixity to which the photographic print, or *épreuve*, stands proof, *preuve*, thanks to the persistence of a perceptiveness that is, for us, the equivalent of the secular city of past centuries.

If taking your bearings is essential in any kind of navigation, the focusing involved in taking a shot is, in fact, even more essential for representation, the setting up of different 'figures'. That is what it actually is, this PHOTOSCOPY that no longer speaks its name.

Take the example of Nadar's aerial photographs taken from a balloon over Paris, not far from the Arc de Triomphe, in 1858. What was that, if not a prelude to this GEOPHOTOGRAPHY covering the terrestrial globe by observation satellites that spy, without letup, on the activity of the planet?

In any case, we might also note in passing that, since Eugène Atget, the image of the city has evolved a lot. Once, perspective views most often represented streets, avenues, the squares in front of cathedrals or palaces and triumphal gates. But from that moment on, the aerial view dominates. And this vision offered by overflight of the excessiveness of the built environment both signals the scale of population concentration and takes it as a target for ever-possible destruction, by firestorm . . .

It is, in short, a 'reduced-scale model' of an urbanism in which the reality of the city seems to be abandoned, since, from so high and from so far away, activity disappears. Here again, the reference to Eugène Atget is obvious in people's lack of visibility, in their missing motion, so characteristic of the photographer's attraction to fixity: the photoscopic exposure that forces itself on him.

Unless it's already about maximum worker mobilisation, the PHOTOFINISH of an industrial race that leads nowhere, as though the film, like the highway after it, demanded an emergency lane to avoid the predictable catastrophe of a pile-up.

Fixity of the geometer's cadastre, framing, or *encadrement*, of 'point of view', the organisation of contemporary man's living environment, or *'cadre de vie'*, demands the polarisation of common perception, the inertia of a moment that passes while remaining memorable – or not. How can we have the audacity to go on pretending that fixity is cadaverous? Looking at it that way, all 'status' as well as all 'statuary' would be mortiferous, even funerary.

As the ultimate anecdote about the TELESCOPIC IMPULSE of the twenty-first century, what about Google's decision 'to make the whole Earth just a click away' for internauts? Thanks to Google Earth, for instance, every one of us can now zap from site to site, amble around weightlessly just by staring at the computer and make out objects scarcely bigger than a standard metre.

'The camera is a lot more than a recording device. It is a
MEDIUM through which messages reach us from another
world,' trilled Orson Welles, as quoted by the photographer
Bill Brandt. With Google, that world is no longer eccentric.
It has become hypercentric, since its telescopy brings us 'face
to face' with the greatest object in history: this *objectum* that is
nothing less than our life-giving star.

The extreme threat of such a 'face-to-face' is obvious to
everyone, with the exception of those who now refuse to
see or have already lost their sight in an OVEREXPOSURE
to the world that deprives them of objective discrimination,
favouring instead the optically correct stance of an instantaneous teleobjective perception of the globe that will impose
itself on us to our doom. This MEGALOSCOPY now brings
all its 'relief' to the old MEGALOMANIA.

Loss of sight, certainly; but, even more, loss of memory,
loss of those 'static shots' of immediate awareness that Susan
Sontag was talking about. After they'd performed the very
first graft of a woman's face, the surgeons said of their patient:
'She saw her face.' Or, more exactly, the face of another
woman who, they say, looked a lot like her. Today, the same
goes for the hidden face of the Earth: we are going to see its
face or, more precisely, its surface, the way we once saw the
surface of the Moon. But it won't be the 'face' of the celestial object that sweeps us along in its course, only the one
produced by the art of a search engine named Google Earth,
whose telescopic impulse is forced on all of us, in lieu of perspective imagination.

What's more, as John Hanke, the director in charge of the
Google Earth product line, indicates: 'The initial idea was
to combine video games with photos of the planet. But our
software was the first to offer "consumer" access to the terrestrial globe from a computer. We wanted, in fact, to satisfy
people's desire to have this interactive planetary experience.'
And so the eye of the master has turned into the eye of each

and every one of us and its megaloscopy, into a mere game like any other.

Almost a century ago, General Lionel Chassin said: 'The fact that the Earth is round has never been taken into account by the military.' Today, not only is this a done deal thanks to the strategists of 'Star Wars'. Those strategists inspired a small Silicon Valley video games company called Keyhole – a name inspired by a generation of spy satellites – to launch itself into digital three-dimensional cartography in 2001. In 2005 the company came up with the famous Google Earth search engine . . .

But all this seems too surrealist to be true. Just as 'the Earth does not move', for its inhabitants it remains, if not exactly flat, then at least horizontal, the spherical nature of geomorphology escaping customary perception. So much so that we civilians are driven to turn around the TOPOSCOPIC question for the military and to ask instead: 'Can the fact that the Earth is flat and fixed still seriously be taken into account by the political decision-makers of the world's nations?' These presidents of a CYBERWORLD that's on the brink of becoming a NANOWORLD, reduced to nothing, or close to it, by the acceleration of an interactive reality in which temporal data compression and the 'dromospheric' pressure of hypersonic transport oppresses us to the point of modifying our mores and 'civilising' customs the same way 'atmospheric' pressure modifies the climate. This is a *geosynchronic* issue that no longer only concerns the fixity of geostationary statellites but indeed all earthlings, *here below*!

Since the memory of history is in fact made up of the static shots of the landscaped PLANIMETRY of geography, let's not forget that the Earth's rotundity is only a PERISCOPIC remoteness, and that what is happening now is a first step in an escape that will shortly be boosted by the escape velocity of rockets, one day or other turning our life-giving star into a point – a more or less brilliant dot, like the dot of the night star.

Let's stop omitting the fact that the telescope is only an instrument of 'delocalisation' and that astronomy has turned the study of the night sky into a hypnotic science, assiduous contemplation of a luminous dot being like talking in your sleep. It ties the astronomer to his optical instrument through the same need for direction that he shares with the hypnotised subject. In the end, that need leads him to attribute to its very 'long-sightedness' the role once granted to the spiritual director!

Over now to one of the most popular astronomers of the nineteenth century, Camille Flammarion:

> A system of deception employed as a means of governance must have requisitioned not only the skill of the scientists of the day, but indeed a whole host of accessories calculated to stun and confound judgement, to dazzle the senses and cause the particular imposture that one wished to establish finally to prevail.[4]

Surely we can't fail to recognize the whole array of our modern telecommunications tools and their 'miserable miracle' in this description. From this eccentric, and soon to be hypercentric, point of view, digital television enjoys not only a hypnotic power over the 'receptive brains of the public' but also a telepathic power over the millions of individuals who make up a 'captive' audience and who are now all hooked, tethered by the chains – the channels – of interactive information technology, thanks to the feats of the multiple screens of our telecommunicating societies.

To top it off, just to come back to Google Earth and its geosynchronous searching for a moment, note that virtual travellers can now meet up on the Google Earth Community website, where they exchange their proper addresses and locations (place marks).

The prospects for tourism of this sort of geostationary ambulation are then multiplied by the number of customisable itineraries. As we might half suspect, the traceability offered to everyone merrily signing up for this 'treasure hunt' works both ways. The choice of their trajectories – their need for direction, as the spiritual director would say – is likely to turn against them, as soon as the tourism market requires it.

Object, subject, trajectory . . . From now on, the trajectivity of waves acquires a crowd-control power that everywhere overpowers the objectivity of any common sense, as well as the subjectivity of this living-present Husserl talked about. Husserl was the man for whom objective presence was still bound up with the object placed before him, as TELE-OBJECTIVITY had not yet deported our immediate knowledge to the ultraworld of the otherworld, beyond the horizon of perceptible appearances, by means of LIVE COVERAGE . . . this 'direct coverage' of the terrestrial globe that suddenly turns any televised representation into a too-obvious PRESENTATION of an allegedly 'whole' world.

Are we children of the stars or children of the telescope? Do we have to choose, or simply submit to what is a purely astrological assertion? According to a number of the world's astrophysicists, man would have to be star begotten and not begotten of the HUMUS . . . even if such a poetic notion stems from the title of a science-fiction novel by H. G. Wells that came out in 1937.[5] It's impossible not to note in passing that, if the stars and their constellations have, as it would seem, determined the origins of humanity in lieu of some 'creator', this is the point where astrology reaches its academic apotheosis, and, after that, there's no point in our being scandalised when some 'card-reading fortune teller' submits a thesis at the Sorbonne!

In the century of *L'Astronomie populaire*, published by the young publishing house of Ernest Flammarion, Camille's brother, Gustave Flaubert wrote: 'The better telescopes become, the more stars there will be.' What Flaubert didn't dare imagine is that such an explosion in outer space would one day take the place of God.

This combined loss of sight and loss of memory is something Gaston Bachelard clearly foresaw, when he wrote: 'Every image is fated to be magnified.' We need to take another look at photosensitive inertia, at the age-old fixity we seem to condemn when we crow about television for having no blind spots; for the emergence of PHOTOSYNTHESIS was one of the wonders of 'biochemical' discoveries . . .

In the course of the 1920s, Edward Steichen decided to try an experiment worthy of Niépce: he put three apples under a thick tent, designed to filter as much daylight as possible, for a record exposure time of thirty-six hours. The result was startling. What expanded was the *relief* of the apples, of each of the apples as well as all of them together. In this still life, then, it is not the depth of field that increases any more, but the volume of the *objectum*, that grows, visibly, to the point of attaining this 'life-size' scale which is also its limit.

Here, the photo-taking apparatus serves first to see the finiteness of the forbidden fruit. Thanks to the time depth of the exposure, what emerges is density – that 'relief of time', passing time, which is, in the end, quite different from the relief of the third dimension of perspective space.

In Steichen's picture, the long exposure length is equivalent to the long focal length of the telescope; so much so that this 36-hour exposure is analogous, all in all, to the exposure of the 'life-size' nature of the memory of the COSMOS. In the end, it is on the basis of such photoscopic focus that we will be able to produce photosynthesis of the astrophysical CONTINUUM – in other words, of space–time from the 'creation' of the world right up to our day. Some people call this the BIG BANG, omitting, in so doing, the intuition of a little-known astrophysicist, the abbé Lemaître, about the primordial atom!

'Light is the name for the shadow of the living light', observed Bernard de Clairvaux. Bearing witness to this light born of the light of the 'Completely Other', the THEOSCOPIC visionary also paid involuntary tribute to the photosensitivity of the historical fact: the very ancient exposure of the image of a Universe that reaches us, after a bit of a time lag, thanks to the shadow cast by light-years that enable us to contemplate not only the stars, those 'maternity wards of humanity', but the very origins of the galaxies, in the revelatory bath of cosmic history.

Actually, if the first screen is the Moon (the night star), the very first imagery of the Universe is the speed of the film of light that enables us to see the history of the past as composed of constellations – and does so without forgetting the cosmetic MORPHING that clouds the perception offered by the lenses of our famous telescopes. I'm talking about gravitational mirages and other black holes that distort the reality of the starry firmament.[6]

There it is, this loss of sight afflicting astronomers and other 'cosmographers' . . . unless what we're seeing as COSMOGRAPHY is actually the intra-uterine staging of an ECHOGRAPHY– in a bid, finally, to catch sight of what is hiding at the heart of the BIG BANG and so tomorrow avoid the involuntary abortion of the BIG CRUNCH!

'Nothing in the universe is fixed', decreed Albert Einstein, as you'll recall. Nothing, that is, except the absoluteness of the speed of light in a vacuum . . . A fine tribute to the PHOTOSENSITIVITY of an instrumental perception that allows us to perform the theoretical PHOTOSYNTHESIS involved in the film of cosmic history . . .

Recently, a documentary entitled precisely *In Utero*, featured a number of gynaecologists attempting to show televiewers what happens during the nine months of gestation.

Once again, what they were bringing out – or putting 'into relief' – with the aid of 3-D medical imagery, was not so much the body being born as the speed of its development: some two million neurons a minute! And so we were able to learn that, from the fifth week of pregnancy, the embryo already has a heart, a liver, kidneys and a stomach; that at the end of the first term, each of its limbs has formed and it even starts to open its eyes. But it is only after the sixth month of gestation that its senses appear to truly begin to develop.

And so we note yet again that, in this telefilm, as so often elsewhere, it's not so much life that is at issue as the speed at which it is propagated; not so much the enigma of the living being as the gynaecological acceleration of its reality.

Chapter 4

The moment is uninhabitable, like the future.

Octavio Paz

'Companions, anti-authoritarians, humans, we don't have much time left', noted Rudi Dutschke just before the attack, in 1968, that was to put an end to his action and, a few years later, to his life . . . That was at the end of the 1960s, those 'swinging sixties' that took us for such a rocky ride, here and there – in any case, far from the reality of the moment: a reality that involved the landing on the Moon and the progressivist illusionism whereby the conquest of astrophysical space marked the pinnacle of an historical time decidedly too *Down-to-Earth.*

In the same period, in the United States John Steinbeck was showing more perceptiveness than Jack Kerouac when he wrote: 'If one has driven a car over many years, as I have, nearly all reactions have become automatic. One does not think about what to do. Nearly all the driving technique is deeply buried in a machine-like unconscious.' He noted the effects of this on a certain category of driver: 'They are a breed set apart from the life around them, the long-distance truckers.' He himself 'had avoided the great high-speed slashes of concrete and tar called "thruways", or "super-highways"'. Forced to take one such road, US90, he goes on:

The minimum speed on this road was greater than any I had previously driven . . . Instructions screamed at me from the road once: 'Do not stop! No stopping. Maintain speed'. Trucks as long as freighters went roaring by, delivering a wind like the blow of a fist. These great roads are wonderful for moving goods but not for inspection of a countryside. You are bound to the wheel and your eyes to the car ahead and to the rear-view mirror for the car behind and the side mirror for the car or truck about to pass, and at the same time you must read all the signs for fear you may miss some instructions or orders. No roadside stands selling squash juice, no antique stores, no farm products or factory outlets. When we get these thruways across the whole country, as we will and must, it will be possible to drive from New York to California without seeing a single thing.[1]

There it is, this 'dromoscopic' bankruptcy, the blindness, the loss of vision that propels the American writer to conclude: 'Localness is not gone but it is going.' For 'no region can hold out for long against the highway, the high-tension line, and the national television'.

It's no longer a matter here of landscapes vanishing into the truck or car windscreen. It's a matter of the disappearance of a territory, of a continent. The aesthetics of disappearance no longer defines film sequences alone; it also affects the art of seeing, or perceiving, ambient reality.

In a recent interview at the University of La Rochelle, the historian Carlo Ginsburg said of his work: 'I prefer to trust reality. Its imagination is more powerful than mine, and then, again, what is "history" if not a fiction that can be proved?'

And so, Ginsburg recognises that, following the acceleration of 'historical fiction' announced by Daniel Halévy in 1947, the time has finally come for this 'reality' that is merely the hyper-power of the Last Empire: the Empire of Speed, which has now succeeded the empire of the wealth of nations, ancient as that was. For the capitalism behind historical ACCUMULATION has now bowed out, economically speaking, before the 'turbo-capitalism' of information ACCELERATION.

In this postmodern eschatology, telereality's imagination puts an end to perceptual objectivity. Once that happens, no longer believing your eyes is not a manifestation of stupour anymore, but a form of blinkeredness, a blindness about the perspective provoked by the light of the speed of an alleged reality that blankets eyesight, any gaze trained on matter. In such conditions, it is not surprising that twentieth-century science, and specifically physics, triumphantly hit on antimatter and its explosive power – in the person of Paul Dirac.

'What was originally evidence has become a postulate', noted Edmund Husserl in 1937. Some ten years later, with Halévy, the philosophical postulate of phenomenonology turned into a sort of ICON, distancing us all the more from this PERCEPTUAL FAITH in the evidence of our sight that the phenomenologist was telling us about. Aware, as early as 1933, of the importance of the future electronic television, its inventor, Vladimir Zworykin, actually gave it the name of 'ICONOSCOPE'!

And so, following Nadar's first 'aerostatic' snapshots of 1858, perception of the world's reality was to give rise to this 'astrostatic' Large-Scale Optics, this sudden ice age of the 'freeze frame': Scan Freeze, the frozen sequence of the perspective of ubiquity that was shortly to subvert the perspective of the space of common reality.

'Terror is the realisation of the law of movement', wrote Hannah Arendt, in the wake of the *Blitzkrieg* . . . thereby illustrating perfectly the very particular fear produced by a terrorist acceleration of History which was to cause so much suffering in the twentieth century. 'Its chief aim', Arendt went on, 'is to make it possible for the force of nature or of history to race freely through mankind, unhindered by any spontaneous human action. [. . .] As such, terror seeks to "stabilise" men in order to liberate the forces of nature or history.'[2]

The keyword here is indeed STABILISE, meaning render inert, or as good as. In fact, the total mobilisation touched off by Futurism and of Nazism was to be followed by the BALANCE OF TERROR between East and West, with the fear produced by an all-out threat, whereby the raging of the arms race would end up flowing into the panic of a world deprived of any emergency exit from then on.

Let's now take a closer look at this 'society of accelerating realism' as it goes into training. Ours is not yet a completely on-line society, but one where entering the virtual community is compulsory, or very nearly, and this means living in a surrogate reality that deprives us of the tactility, the physical contact and the empathy essential to communal intersubjectivity.

Today 56 per cent of adolescents opt for chat rooms to maintain contact with their acquaintances. On the other hand, only 51 per cent of those who form relationships over the internet actually take the plunge and meet their interlocuters in the flesh. As a result the firms Google, Yahoo and AOL have made this kind of instant messaging service available to them.

So proximity without promiscuity has become a market. Proximity stops meaning 'here' and turns into 'OVER THERE'. Objectivity *de visu* is suddenly transformed into the (tele-audiovisual) objectivity of a lifeless tête-à-tête, a face-to-face via an INTERFACE – only, with an unobstructed view of screens that stand in for the shared horizon.

What is at stake here is indeed loss of sight in an 'art of seeing' whereby the gradual narrowing of the visual field to the frame of the screen (of all the screens) is the perfect equivalent of this eye disease that reduces our lateral vision and that goes by the name of 'glaucoma'. Glaucoma is an irreversible disease, most often painless, that causes a number of optical data to disappear one by one and that can develop into total blindness. A SHUTTER OVER OUR EYESIGHT, glaucoma achieves a sort of furtive iconoclasm not of pictorial imagery any more, but of objective ocular imagery. It thereby affects our mental imagery and so, our subjectivity – all this to the exclusive advantage of an instrumental, teleobjective imagery which, by distancing us accordingly, interferes with the intersubjectivity of contact between interlocutors. That is the very definition of a twenty-first-century TELEREALITY that can't be summed up by the party games of 'commercial' television stations.

As Maurice Merleau-Ponty put it so well: 'After all, the world is around me, not in front of me; I inhabit it.'[3] To suppress, as does the shutter of the screen of a so aptly named 'terminal', not only lateral vision, but also the countryside, the land around, is to deprive all customary reality of relief and to experience a disastrous reversibility of dimensions, in particular of the depth of perspective.

'There is no distance any more, we are so close to things, they no longer concern us at all.'[4] When that happens, we navigate blind, just like Steinbeck's thruway drivers, no longer on the surface of a nation now but on the surface of a closed world; a world FORECLOSED by the 'temporal compression' of a 'realist' acceleration that is nothing less than the realisation of this totalitarian law of movement now known as OUTSOURCING.

Far from being 'in the world', as in days gone by, we are now only at the edge of the world, at its very last extremity . . . And that is exactly what is causing the current reversal of the notions of exterior and interior, outside and inside, with the reversibility of dimensions.

In fact, with the global system of instantaneous telecommunications taking over from the totalitarian systems of the era of general mobilisation of transport – or, if you prefer, of 'the attack on the world' (Heidegger) – 'the global means the inside of a finite world, and the local, the outside; in other words, the great suburban outskirts', all that can still be precisely located here or there.

Not so much topological as 'toposcopical', the sudden reversal in our relationship to the world around us – a result of the acceleration of a real time that dominates shared space on all sides – demands training, formation, a sort of 'tele-scopic' education. And this is required from the moment the child 'comes into the world', earliest with infancy now accompanied not so much by the child's biological parents as by the screen relations that will so tightly surround him as an adolescent and as a grown-up in days to come.

Designed for the under-threes, the very first television channel for babies saw the light of day (!) in Europe, in the autumn of 2006, after having been launched in Israel for Christmas 2003 . . . Baby TV is targeted, you could say, at the telesubjectivity of earliest childhood, in prime time, using short two-to-ten minute segments, and a particularly slow pace aimed at assisting their absorption as much as possible. Offering cartoons, nursery rhymes, games and documentaries, this channel pushing desensitisation to and embrace of the acceleration of reality trains the baby in the 'optically correct' perception that will evolve into the aesthetics of his years as a grownup. It thereby promotes toddlers' addiction to the small screen, at the same time as it claims to be protecting those toddlers by drawing the parents' attention to the risks of their progeniture's habituation to the specialised channel's 'hypnotic effects' . . .

It is recommended, for instance, that the mother accompany her child with appropriate commentary and avoid at least bottle-feeding, if not breast-feeding, while a show is on, so the baby won't identify the imagery with feeding . . . Unless the point is quite simply to avoid the child's swallowing this alternative mother's milk the wrong way!

Such advertising duplicity leaves a person dumbfounded. Whereas 'corruption of a minor' is a practice universally condemned, this toposcopic corruption of the relationship between mother and child is literally confounding!

As one child psychiatrist points out:

From birth to eighteen months, the infant learns to recognise its own image and to apprehend reality, and this is so that it can detach itself little by little from the body of its mother. If, from the first months, it is offered another dimension, *this one virtual*, doubt may dawn in the infant and scramble its initial reference points. Whence the essential role parents play in helping it to distinguish what is real from what is virtual.

The childhood psychiatrist doesn't seem to understand that this is precisely what is happening today: this very distinction is being scrambled in order to favour the infant's future OVEREXPOSURE, once they've reached adolescence, then adulthood, to the telereality of a now 'closed-circuit' world. Which also explains the broadcasting of Baby TV twenty-four hours a day, even though parents are strongly advised not to leave very young children in front of the TV, at night . . .

Actually, with this programme aimed at desensitising the child to loss of a sense of reality, we find ourselves faced with a deterrence machine that has no precedent, even though it basically carries on the role of telepathy and the craze that excited in the 1930s, particularly in Austria, as the Austro-Hungarian Empire came to an end. In those days, Europe practised collective suggestion on the masses and a form of 'hypnosis at a distance' to disastrous effect, culminating in the tragedy we are all familiar with . . . but on such a change of scale that the drama evoked seems today to be truly minor.

In fact, as we've seen previously, with that network of networks, the internet, all the dangers of the outside world are invited inside, into each and everyone's home; for the TOPOSCOPIC reversal is taken to the dizzy limit in – and all over – the internaut's abode, if not in the palm of their hand, thanks to mobile phones equipped with video teleconferencing.

Here, the ICONOCLASM of the ocular image takes on the proportions, if not of a 'crime against humanity', then at least of a sort of torture (at a distance) of the visual field

and of the affects of childhood, whereby loss of a sense of
reality turns into this 'business of appearances' (I talked about
some twenty years ago already) designed to favour the future
GLOBALISATION OF AFFECTS, thanks to the sudden
synchronisation of emotions.

A lover of popular and quasi–populist astronomy, which
he enthusiastically promoted, Camille Flammarion was also
president of the Société aérostatique de France and, like
Nadar, a devotee of hot–air balloons. Way ahead of the
Italian Futurists, Flammarion may be considered the prophet
of this society of accelerating reality which was to lead to the
on–line society of instantaneous telecommunications in the
twentieth century and especially in the twenty–first.

Aware of the fact that astronomy is a hypnotic science
due to prolonged observation of objects shining away in the
firmament, Flammarion was to take part in Charcot's experiments at La Salpêtrière hospital, in particular those to do
with the cataleptic effects of electric light on the celebrated
neurologist's patients. Similarly, only in Le Havre this time,
he would be a privileged witness to the experiments of Pierre
Janet on phenomena involving mental suggestion, which
Janet himself described in his bulletin of 1896 as 'a som-
nambulistic passion that binds the patient to the hypnotiser,
to the point where the hypnotized person's need for direction brings him *to attribute to his therapist the role of a spiritual
director.*'[5] That was the ultimate accolade for the progressive
atheist savants of the day!

But let's hear what Flammarion had to say on the subject
of telepathy: 'It seems to me that it is a matter of trans-
mitting images by psychic waves between two brains in
harmony, *the one fulfilling the role of wave-emitting device and
the other, of receptor.*'[6]

'Telescopic' hypnosis, somnambulism, catalepsy and,
finally, telepathy: the height of mental suggestion would in

the end be achieved by this telereality accelerated by the light of the speed of waves which, they say, 'electromagnetise' our receptive consciousness. For some, we could even talk of a reality enhanced by cybernetics . . .

By way of confirmation of the *prenatal* character of contemporary practices in the CYBERWORLD, let's turn now to Auguste de Frarière, the author of *L'Education antérieure*, or *Earlier Education*, published in 1855: 'As though by suggestion, a somnambulist is likely to perform changes in his organism; for a still distant future age, we can safely assume that what a pregnant woman can do on herself, she can also do in the same way on her child.'

Like some 'sensation incubator' the company Life-Style has just put a product called OCULAS on the domestic market. This is a space capsule designed to remain on Earth in which the internaut enjoys perfect instrumental occlusion of the ambient world in order to give free rein to their telescopic passion, thanks to various programmers of an audiovisual telepathy that allows the internaut to absent themselves from the shared world – or rather to BE BORN THERE FOR NO ONE!

Better still, to get away from the stress of realist acceleration, the French firm Cocoon's has put photosensitive inertia into (neutral) gear, with small individual cabins – Wellness Cocoons – where treatments of all kinds are dispensed. The incarcerated user sets themselves up inside a sort of bubble where a 'smart seat' scans close to sixty points over their body and offers them a personalised vibratory massage, while the cabin broadcasts relaxing sounds, like the backwash of the ocean or a gently sighing breeze, and images stream past through electro-luminescent goggles.

After the sensory deprivation tanks once used to deliver leisure to the agitated or torture to troublesome political detainees,[7] the hyper-realist incubator prepares us for the birth under the letter 'X', not of the 'Eve of the future', but of a being Villiers de L'Isle-Adam could never have foreseen!

But let's get back to the relationship of dependence between hypnotiser and hypnotised that allows the former to become his patient's 'spiritual director'.

Today, in Great Britain as in Italy, insurance companies force car drivers who have held a driver's licence for less then three years to equip their vehicle with a 'black box'. This is connected to a GPS satellite surveillance system which ensures that their course on the road is monitored, in real time, thanks to speed readings every two minutes, checking the duration of the drive along with the itineries taken. Thanks to this trajectography, the novice driver finds themselves in the position of a detainee conditionally released and condemned to wear a TRANSPONDER, that 'antiquated capsule' of days gone by!

It was only one small step from the mollycoddling inertia of the internaut encapsulated in networks to the endless GEOLOCALISATION of the externaut's freedom of movement: 'One small step for man but a giant leap for mankind' . . .

The reversibility of dimensions is evident here again. We note, too, that the TELESURVEILLANCE of the various events taking place everywhere within the space of globalisation is also accompanied, but on the sly, by this incessant GEOSURVEILLANCE of our customary displacements. Some people are calling this the traceability of the whole set of the subject's trajectories. With it, our private geography is suddenly split, turning into this point-by-point monitoring, this trajectography that allows our *activity* and our *interactivity* to be surveyed at all times.

Already the most spied-on citizens in the world, the British will soon be subjected to an automatic numberplate recognition system which will film the movements of all their different vehicles. Endlessly analysed by a national data bank, this information could be compared to that collected by customs or the police. Lastly, we might remember that the installation of telesurveillance has made remarkable progress in England, with one million video cameras in 1999, three million in 2003, twenty-five million by 2007. So extensive is it that an innocent Londoner is now filmed *three hundred times a day* by different video networks . . .

Remember that, since the terrorist attacks of the summer of 2005, the approval rate of these inquisitorial measures sits at around 70 per cent. Anyone would think the terrorists had already won the offensive against civil rights.

In his famous interview with Paul Gsell, Rodin said of Niépce's invention: 'Photography lies for, in reality, time does not stand still.'[8] Not only was the sculptor of 'walking man' right, but he anticipated the numerous aesthetic problems of the megaloscopic age.

Indeed, not only does time not stand still, in fact, but it seems today to speed up to the point where, thanks to satellites, it reaches the 'dromospheric' limit of the globe. We are now at a stage where we might in turn paraphrase Rodin and say that: real-time television lies for, in reality, space does not stop at the limits of screens any more than it stops at the limits of terrestrial rotundity.

Despite the advocates of the market economy of proximity, the Earth is clearly not flat, far from it; it has contracted, like a muscle suffering from a painful cramp. That is all it is, this GLOBALISATION of the world-village. Contrary to what 'virtualists' of every stripe may say, there is therefore no 'enhanced reality', but only a Real sped up to the point of ushering in the ethological disasters to come.

In this sense, the deterrence evoked in the preceding chapters, far from having been an enhancement of 'historical reality', would merely have been its fatal restriction; whence the hidden threat of an accident in knowledge acquired over the ages.

Let's hear what Thomas Friedman, that flat-world Columbus, has to say: 'Whatever can be done will be done and a lot faster

than you think'.[9] Impervious, it would seem, to the irony of
what he's saying, the American columnist is thrilled by the
radically unfathomable and unprecedented nature of the new
market in interactive proximity and openly admits to being
guilty of a 'technological determinism' that leads him to define
our telluric planet as a technological platform.

Taking up Karl Kraus's phrase about the ravages of psy-
choanalysis, Friedman goes as far as considering that 'the flat-
tening of the world is both the disease and its cure'. He even
cites an analyst who claims that 'permanent' jobs are not so
much outsourced within the space of relocation (to India, to
China) as outsourced to the past, once they are digitised and
automated . . .

As we might guess, this programmed disappearance of
time-honoured trades doesn't just lead to the ruins of the
past, but to those of the future; and, if idleness was once considered the mother of all vices, structural unemployment is,
for its part, the father of all the physical abuses to come!

In relation to this state of play, where the world time
of astronomy seems to dominate the local time of urban
agglomerations and their exchange activities, a comparison
needs to be made. We need to compare the non-separability
of the infinitely 'small' in the realm of quantum mechanics
with what happens when it is transposed to the infinitely 'big'
of the non-separability of the media realm.

With all the confusion in feelings of belonging and with the
drift of the five continents that make up geographical space
towards the sixth continent of cyberspace, suddenly the mor-
phological stability of reality is threatened with collapse. If it
goes down, it will not only drag culture down with it, but also
– equally – the most durable reality there is: the reality of the
orientation, not of some 'hypnotic' vision now as in the past,
but of the very fact of being-in-the-world and the rationality
that goes with it.

On the threshhold of the conquest of space, Wernher von
Braun, the Nazi rocket developer employed after the war by

NASA, noted: 'Tomorrow, learning space will be as useful as learning to drive'. Marking the difference in kind between the revolution in 'astronautic' transportation and the revolution in 'cybernetic' transmission, we can all too easily divine the traumatism that awaits us when we are faced with this TERRA INCOGNITA thrown up by a spatiotemporal exile, in which the loss of the art of seeing and of knowing belonging to phenomenology, will be doubled by the loss of the art of conceiving esential to our *being there*.

'Time is the cycle of light', wrote the theologian Dietrich Bonhoeffer. But this cyclical time of seasons and days is now doubled by the real time of an instantaneity that is merely the cycle of the speed of light waves which convey the information of image and sound.

What then builds is the 'dromospheric' pressure that largely changes the realist climate of our days and years, and with it, the stability necessary to our perception of the world, along with our perception of our strictly human identity.

PRESENT in the world, then TELE-PRESENT thanks to 'geolocalisation', the trajectory is imposed on the subject the same as on the object in a just-in-time supply chain that now dominates the time-honoured stability of product stocks.

Imperceptibly, communication at a distance morphs into tele-communication before mutating into pure PROMOTION, in order to PROMOTE everyone's displacement, proximity thereby translating into the hypermarket of the future . . .

But this campaign promoting physical movement is still not enough. It has to be topped up by the promotion of our emotions – to the point where, alongside the electronic intelligence mode, the I-MODE of the computer, they're now talking about the necessity of I-MOTION due to the high flows involved in information technology.

So it's probable that tomorrow, with the looming climate catastrophe, learning the lapse of time remaining will be just as useful as learning to drive a computer!

Some concrete examples reach us from the land of the Rising Sun, where OTAKU culture has developed in tandem with manga art: in the big Japanese retail centres, customers today seem to be guided by a sort of *cell compass*. Intrusion into the private lives of Japanese citizens is no longer the privilege of some sort of political police; it is the privilege of an insurance police attached to a massive retail operation in quest of buyers. Whence this strictly totalitarian temptation to use crowd phenomena as a new kind of mass media, one based on the power to control a population that is no longer static and sedentary, as it was in the not so distant past, but permanently mobilised for whatever advertising or ideological motive. This development is along the lines of what already happened twenty or so years ago with 'just-in-time' mass distribution using global positioning satellites for vehicles transporting goods.

We can now better understand the development of the mobile phone, its sophistication, while we anticipate some day soon wearing intelligent garments, those future 'electronic straitjackets' that will complete the 'chemical strait-jackets' once used on detainees, Soviet or otherwise . . .

It is all about paving the way for a universal remote control that will no longer look so much like the remote control involved in instantaneous telephonic virtuality as like the kind involved in the vitality behind being-there, here and now. In such a remote-controlled existence, the individual will be kept in constant contact, at every moment and at every point in their trajectory, so that they will no longer be left with any spare time. In other words, any free time, time for reflexion, for prolonged introspection. For tomorrow we will all be monopolized by the growing outsourcing of our once immediate sensations; we will all suddenly be collectivised in our affects, in our most intimate emotions, slipping and sliding or, more precisely, 'surfing' as we will then be in a new sort of epidemic of cooperation; the pandemic of a mob once solitary, now plagued with the delirium of a UNANIMISM that the prophets of doom of the twentieth century foretold.

This will happen thanks to the perfect synchronisation of once fleeting emotions that will, if we're not careful, take over from mastery of our feelings and opinions (political or otherwise); for the *freedom of expression* offered by telecommunication companies will suppress the *freedom of impression* of a by then captive audience.

'It's too late to have a private life.' That phrase from Joseph Losey's film *The Damned* beautifully illustrates the coming of those intelligent hordes announced by Howard Rheingold in his essay, so judiciously entitled *Smart Mobs,* where what is at issue is precisely the social revolution in electronic cooperation: in other words, in the very latest form of INQUISITION, brought to you by the people's tribunal in the age of general interactivity. A form of sorcery, of electromagnetic bewitchment, electronic cooperation would take on all the excesses of the extralucid telepathy we started out with millennia ago under the fallacious pretext of a collective simultaneous intelligence that could turn humanity as a social corps not into 'a single people for a single Führer' this time round, but into a single MASS MEDIA CORPUS.

In this unprecedented mysticism of a communism of affects, the masses would be controlled by telepathic means, point by point, individuality by individuality. Such a process that would thereby achieve the perfection of a mass individualism that would intensify the ravages of the collectivism of days gone by. Messages from the beyond would no longer be the ones relayed by the religions of monotheism; they would be the messages generated by this sudden outsourcing *hic et nunc* of each and every one of us. And as if by magic, the result would be a sort of MONOATHEISM – not based on people's freedom of conscience anymore, but on the eccentric discrimination of a 'Global Brain', representing humanity and inspired by certain theological theses of the past century –

only, stripped of Christ. It would, on the other hand, be lumbered with the electromagnetic ghost of a 'Great Computer' able to automate, not the formation of some mystical body of the human race any more, but a media body unified by the merits of instantaneous interactivity.

This is a fusional form of hypersocialism, whose metaphysical outrages are everywhere present or, more exactly, telepresent, in the era of single-market turbocapitalism and the singleminded so-called 'thinking' that informs it, that new political orthodoxy driven by free-market principles.

There was a time when you could say of a *psychotic individual* that he was 'losing his mind'. These days we should say of the *exotic individual* produced by the mass individualism now taking over from totalitarian collectivism that he is 'losing his World'!

And so, the acceleration of the real time of telepresence confirms the old adage: 'If you want to go fast, go alone, but if you want to go far, go with others.' 'You've put working people in college', Georges Bernanos lamented apropos the tyrannies of his time. Now, we could paraphrase Bernanos and retort: 'You're putting the world on the switchboard!'

Here they are, those 'smart mobs' galvanised by a general interactivity whereby the *watchdog* of moribund communism has suddenly turned into the *seeing-eye dog* of hyperliberalism for an idle humanity whose handicap is carried to the dizzy limit by the straitjacket of the radiant future. This 'smart garment' smart enough for two will free our hands of the cumbersome 'portable', with 'remote control' of our customary vitality advantageously replacing the remote control of the old domestic 'TV set', whose last show has long been in the can, in anticipation of the nanotechnologies that will feed the body of humanity with electrotechnical fuel.

Where the transport revolution simply accelerated the history of industrialised societies, the instantaneous transmission revolution is no longer content just to accelerate real-time communications: it's gearing up, as we've just seen, to accelerate the reality of the living body's physiology thanks to a third revolution, the transplant revolution. The pacemaker has only ever been one of the precursors of this new revolution, anticipating dual biotechnologies, since there is nothing in biology that is not transposable when it comes to fitting out bodies for the race of the living being through time.

Today, then, reality is merely a remnant, a residue of the growing progress of the instrument. Yet, just like our vitality, reality is a *gift* not merely some *given* among other similar data. In this sense, and only in this sense, there exists a sort of AUTOCHTHONY OF THE REAL which the development of information propagation speeds puts in peril: that is precisely what this 'postmodern' loss of the sense of reality is.

Teleobjectivity is not a simulation, then, but the on-the-spot contraction of a unity of time without unity of place; we are its natives and it now conditions the vitality proper to each and every one of us.

In this sense, the contraction of the semblance provided by the screen only ever simulates the accident to come, an accident in the self-sufficiency of the solid world. Once, any figurative representation was only a reduction of the space occupied by the forms of the visible. But the presentation of the world in the contemporary age of telepresence looks, for its part, like a muscular contraction that suddenly sends pain spreading through the whole body.

At the start of his book, *Eye and Mind*, Maurice Merleau-Ponty writes this revelatory sentence: 'Science manipulates things and gives up living in them.'

With the technosciences of our cybernetic virtuality, this renunciation becomes a sort of apostasy and concerns everyone who uses its instruments, its tools of perception or conception. And so the 'rites of autochthony' within the real

space of the ancient city of the Greek world are now replaced
by the rites of autochthony within the real time of a so-called
'world village', that tends to resurrect the *AXIS MUNDI* of
the cities of Hellas as the hypercentre of a place that is no-
place, where the connection between the 'citizens of the
world' gradually takes the place of all the places of customary
locality, thereby destroying the very foundations of politics.

In an article in *Sud-Ouest* in 2006, the climate historian
Emmanuel Le Roy Ladurie warned of the political dimension to climate accidents like the heat wave of 2003, by going
back to the source of the French Revolution in the heat
wave of 1788 and ending with cyclone Katrina and the global
warming currently well under way. He thereby publicly
flagged the inaugural nature of the first protest against the
Météorologie nationale, France's national weather bureau,
accused as it was of failing to predict repeat disasters – from
the 'storm of the century' at the end of 1999 right up to that
heat wave of the summer of 2003 and the dismissal of the
Health Minister at the time, Jean-François Mattei.

There is atmospheric pressure on the history and politics of
nations, but the dromospheric pressure of public opinion and
of progress in knowledge on political mores is stronger still.

War or revolution, these collective passions of the day
are dragging the most developed societies down into the
delirium of 'interpretosis' and panic, as though the bad
weather culture whips up already long coexisted with the
bad weather nature provides, thereby wreaking havoc on
the history of civilisations.

And while we're still on the subject of autochthony, we
might just add that there also exist veritable SEASONS OF
THE REAL similar to TRENDS in political economics.
Their history remains to be written, linked as they undoubt-
edly are to the deleterious climate of 'natural' catastrophes,
but also, equally, to 'industrial' and other such disasters.
This is very much what we have been feeling since the

third millennium kicked off, with the 'transpolitical effect' of calamities that affect our CYBERWORLD. The original obviousness of perceptual faith that Husserl was talking about is little by little giving ground (and how!) before the terminal obviousness of a perspective faith in which the telereality of real time will once and for all put paid to the *hic et nunc*.

IV

The University of Disaster

Chapter 5

Once we spread out into space our future should be safe.

Stephen Hawking

Devoid of peripheral vision and attracted by the central region of screens that accompany him in his displacements, his escape routes, the 'constant' televiewer cum mobiviewer drifts not so much off his course these days as off-course in his life.

A DROMOMANIAC and deserter of the environment traversed, this speed addict participates, unwittingly, in the great mutation in planetary settlement whereby the sedentary man is now everywhere at home thanks to his mobile – in the TGV, in the supersonic jet as in his place of residence – and the nomad is nowhere at home, despite emergency refuge centres and the tents that clutter up the footpaths of Paris.

After the great freeway catastrophes involving collision between vehicles travelling in convoys – repeat catastrophes, mostly due to a lapse in attention to what really surrounds the mesmerised driver – the pedestrian, too, has started straying off-course, at the risk of in turn causing accidents related to his 'passing locomobility', and so, behaving like someone in a state of drunkenness.

Manifesting a sort of aleatory approach in this way, unwittingly, the ordinary passerby becomes an involuntary choreographer of a handicap which certain nervous diseases are a patent sign of. Not content to eliminate anything to the side of his path from his field of vision, the contemporary pedestrian of cellular visiophony is so busy concentrating on the audiovisible interlocutor he's calling, he scarcely sees in front of his nose.

Faced with the improbable success of such telephone practices, surely we've worked out that this new 'body technology' will soon give rise, not so much to some kind of censure, as the opposite: to enthusiastic celebration of this kind of postural drift and the disjointed, unbalanced style of walking that goes with it. This is already happening with the evolution of an art of dance in which acrobatic postures have replaced the harmonics of the corps de ballet, with falling even becoming the highpoint in the progress of choreographic feats.

So, making himself unfamiliar with the immediate area lining his walk – just as the car driver does with the verges along his route – the solitary pedestrian will perhaps, one day, wind up completely neglecting the life around him, so close to hand as he rambles along, and become instead completely engrossed in the collective imagination of an audiovisible 'far-away land' that will satisfy his expectations to the detriment of any actual encounter.

We can also imagine the kind of head-on collision or pile-up of those incredibly smart yet solitary mobs Rheingold was on about, mobs incredibly OBJECT-oriented and notoriously SUBJECT-disoriented, with radio-controlled characters bobbing along, one after the other, and crashing into like-minded types they hadn't spotted.

At that point, urban displacement itself would be made 'uninhabitable', on top of any abode, and getting around day by day would become a sort of 'obstacle race' in which the other person ceases to be anything more than an adversary, at best a competitor, one you only encounter once, in the general panic of collective terror. Freedom of movement (the first freedom of the animate being) would little by little give way to the policing of the streets against 'uncivil' behaviour, thereby completing the policing of highways and football stadiums, since the thoughtfulness and courtesy that were

hallmarks of the old urbanity would have completely disappeared, vanished into thin air – thanks to the airwaves, as is
already the case in a number of suburban areas.

We might even assume that, like highways, pedestrian
traffic corridors might be needed: a fast track for young people,
teenagers; a middle-of-the-road track that would be a sort of
compromise for the middle-aged; and finally, not far from the
emergency lane, a senile track for the outmoded old!

By way of confirmation, note that, at the start of 2007,
the Department of Road Safety, on the initiative of the
Gendarmerie nationale, France's national police force, and
of the Department of Education, issued a PEDESTRIAN'S
LICENCE for schoolchildren in the south-west of France.
This licence, they tell us, 'targets children between eight
and nine years of age and is meant to enable them to absorb
the road rules for travelling on foot as well as the automatic
reflexes needed to ensure their safety when out walking in
the street or along the roadside'.[1]

Accordingly, all children of school age will receive a booklet
entitled *The Pedestrian's Licence Code*, which will enable them
to make strides outside school hours. The operation's slogan
is: 'ON FOOT, YOU'RE THE ONE DRIVING'.

As the captain commanding the Gironde police squad
observed:

> Though the number of people killed on the road has gone
> down overall, the number of pedestrians killed has gone
> up. As for the child pedestrian, he's a lot more vulnerable
> because of his small size, which means he has limited vision
> of the dangers of the street. Kids have limited visual acuity,
> and too often they don't judge distances very well. Bringing
> in a 'pedestrian's licence' should help them learn and respect
> the basics.[2]

'In order to create, I destroyed myself; I have externalised
so much of my inner life that even inside I now exist only

externally', Fernando Pessoa noted, illustrating brilliantly the state of panic-stricken dependence characterising mass individualism.[3]

In fact, whereas our displacements were once restricted by trails, paths and country roads, the city multiplied these compulsory routes in its streets, avenues and boulevards and in its public squares. Now, it is our gestures, our slightest movements that are watched, sensed, underscored by the techniques of automatic pursuit employed by all the detectors of our traceability: video cameras, radars and other sensors.

Following the old example of the TRANSPONDER, which writes remotely, on a police monitor screen, the compulsory itineraries of detainees fitted with electronic handcuffs, each and every one of us is now under the control of waves carrying the messages of their 'cell' phone, so aptly named, reminding us of the *fourgon cellulaire*, the police van that is also a cell, one used to transport defendants to court.

From the age-old trajectivity of our long-haul journeys, we have shifted imperceptibly to this passing INTRANQUILLITY of a dromomania in which on-the-spot gesticulating is only an indication of a growing inertia that will not fail, tomorrow, to root the solid world of idle humanity to the spot. Whence the Portuguese poet's sigh: 'Ah! Not to be everybody and everywhere'. To which I for my part would add: 'Everybody and all the time'.

But let's hear more of Pessoa's muted voice: 'God is an immense bridge, but between what and what? As far as I'm concerned, I've lost the plot: I'm just the bridge between me and myself.'

And so, after the loss of the autochthony of a reality that inhabits the Earth, we seem to be wondering, now, about this cosmic INTERWORLD that claims to supplant the INTERVAL of the time distances of our life-giving star, in order to safeguard humanity's future – as Stephen Hawking sees it. While we refrain from praising God 'in the firmament of his power',[4] where the stars break out, the British astrophysicist wants us to break loose with the help of astronautics in order to relaunch the old colonial adventure!

They say that NASA is ginnying up to relaunch its permanent moon base project, in 2020. Now there's a fine example of RELOCATION for you!

Actually, it's not Pessoa who has lost the plot; it's our history that is going into exile up in the wild blue yonder on the way to the next world, in a madness in which science and fiction merge.

What more can you say about this loss of sight produced by a century of inattention, the so-called twentieth, where the 'telescopic' optical illusion was brought to its point of incandescence, 'brighter than a thousand suns'? Whatever they might say about Progress over the past century, when we talk about the MOON, we should also talk about the SUN, the sun of Hiroshima, since the feats of the physics and astrophysics involved in the moonlanding were never anything more than the shadow cast by the Great War of Time, a war that soon turned into Star Wars.

And this is where Nietzsche's famous dancer comes in, out of the mouth of William Forsythe: 'Welcome to what you think you see!' DROMOSCOPY cum DROMOMANIA, the art of choreography is rushing to take back its right of primogeniture over the theatre of language.

An art of the traceability of bodies in movement in which the ballet master has, in advance, all the attributes of a 'spirtual director' (or of a software consultant), dance has already upstaged the visual arts, in the popular imagination, by being a living art. Which amounts to a clinical symptom of an age where trajectivity holds sway, everywhere you turn, over time-honoured objectivity and the subjectivity of the animate being.

Take the example of Trisha Brown at the Paris Opéra.

Under Brown, the dancers hardly move around at all anymore, they just jig a bit or, rather, vibrate, like the electro-luminescent waves that ruffle our screens.

With dancers appearing and disappearing from one wing to the other, the choreographic stage is now a mere thoroughfare where dancers pass more or less furtively, shooting onto the boards *ex abrupto*, only to flit off again three seconds later, as though they'd got the place wrong . . .

In her 2002 piece *Geometry of Quiet,* Trisha Brown refers back to a dance involving exorcism and some sort of bid to cure the evils of the times: 'Surrendering to the pull of the wings and bringing them on-stage, capturing their vibrations through those electric batteries, her dancers', the American choreographer is clearly not fooled by the illusion of real time offered by our hypermodernity.

Another witness to such choreographic dromomania is Odile Duboc. In *Nothing Allows Us to Guess the State of the Water,* the hypnotic trance lasts throughout a show where the lighting shapes the space more than the choreography does. All this begins, as so often, with the running on-the-spot so dear to post modern dance, in which dancers adopt perfectly inert sculptural poses at the end before collapsing in a heap on the ground . . .

So many signs that reveal the anxious anticipation of an era in which language no longer says anything and the word no longer mediates human relationships. Which explains the 'hyperactive' quality but, especially, the return to body language, in a signaletic gesticulation that now everywhere accompanies the dromomania of an individuality freed from any social or territorial tie.

Since the beginning of the twenty-first century, in any case, these vibratory or, more exactly, VIBRATIONIST practices, inspired by the experience of weightlessness and spatial expatriation, have even contaminated home sports for those who still want to stay in shape.

Whether it means riding an exercise bike or doing gymnastics, glued to the box, practising sport at home has become extremely popular. So much so that the cumbersome running-on-the-spot machine that removes the need for jogging is starting to be overtaken as a preferred option by the POWER PLATE. This is a platform that sends the user from thirty to fifty muscular vibrations a second. Invented by the Russians to combat the harmful effects of weightlessness (muscular atrophy, bone degeneration) on cosmonauts manning space stations, such vibrationist training is starting to spread – on Earth, this time; in particular in centres of rehabilitation, convalescence, thalassotherapy and, equally, in the homes of sport and media personalities; the infatuation with fitness now making way for wellness, today's strong trend towards effortless well-being . . .

The great mass movements of the past century's revolution in automobile transport will no longer survive, beyond city and suburban traffic jams, except in this trend towards crazily aimless jigging around such as performed by our 'convulsionary mystics of rhythm'. Not really 'dancers' anymore but *contortionists*, these people distinguish themselves more by their acrobatic feats than by their choreographic art. It's as though the Great Movement of Progress ended, finally, in a type of individual jerked around by the insane swivels and nervous jolts symbolised by hip-hop. It has got to the point where we recently saw a Japanese carmaker use one such on-the-spot dancer to sell its fast cars!

But it's true that in the arena of 'great touring car' automobility and driving, *conduite*, as in bad domestic conduct, *conduite intérieure* (also a saloon or sedan in French), we are currently seeing 'eccentric' practices. In Great Britain, for example, the State Secretary of the Home Office has just introduced a law aimed at coralling prostitutes in apartments, so as to clear the streets of their trade. As for the planned penalty: 'Any marauders on the prowl for sex on public thoroughfares will find their driver's licence suspended.' Obviously, outsourcing is not yet as widespread as they say!

Once, at any town or country dance, people danced in bands, in 'sarabands' and in lines, from the *branle* to the *bourrée* or the *menuet*. Later they danced in couples, changing partners at random. Now, thanks to the 'political incoherence' of single-parent families that accompanies the 'quantum incoherence' of the physics of the infinitely small, detachment is at a peak: now everyone around in quest of a more or less solitary pleasure. And this pleasure is nothing more than a kind of divorce by mutual consent, in these nightclubs that are clearly schools of urbanity, as they are about to open a number of them that will be exclusively reserved for children aged between ten and twelve – anticipating the day when nursery rhymes can be followed by the decibel orgy of the raves that are set up these days on military grounds and disaffected airports.

As for the kind of energy expended around the place, here or there, note that a London-based design firm, Facility Architects, has just come up with the Pacesetters project, which aims to 'harvest' the abundant renewable energy generated by the vibrations of human bodies circulating in railway station hubs and to light railway concourses with it. It will do this by means of a staircase whose risers will contain technology which will pick up kinetic energy from commuter footfalls and convert it into electricity!

'Other people look straight ahead, we look up'. Heroes of an upward dromomania, the French 'speeders' who call themselves Yamakasi practise an art of displacement that signals, before the eyes of all, what I once called falling upwards, in relation to the velocity of escape from weightiness.

Fans of a kind of tightrope walking for city-dwellers in which the 'cat burglar' is no longer a housebreaker but a record-breaking clown, the Yamakasi have an inverted view-

point which, from the lost horizon of distant suburbs to the
zenith of the high-rise towers of the Ile de France housing
estates, illustrates beautifully what an impasse these 'free zones'
are. For the towers are never anything but dead ends at alti-
tude, vertically extending the foreclosure of those neighbourhoods you have to get out of any way you can, including
by taking a 'swallow dive', or by rock-climbing, or by what
might swiftly turn into the death dive of a – final – escape.

Let's not forget that, in the beginning, the DROMO-
MANIAC was not so much a hurdler, a nomad or a madman
as a *deserter*, a man on the run. So it is perfectly revelatory to
see the Yamakasi lavishing their *savoir-faire* on the detainees of
Fleury-Mérogis Prison. The inmates there ask only to be like
them in order to break out, at last, from the enclosure of the
penitentiary, with the place of confinement, like the 'very high
building' (VHB), turning into an architectural structure to scale
in this rock art of unbridled alpinism; a 'playground for on-the-
spot play' which frees itself from the mountain chain, from the
dizzying walls of the Grandes Jorasses or the Dolomites, and
turns instead into the excessive hijinks certain alpinists have
recently initiated. Here, the long mountain treks of yore make
way for solo speed races where the aim is to 'chalk up several
summits in no time at all' – if possible, in the same day. The
prowess of the seasoned alpinist gives way to this climbing
speed where competitors use BASE JUMP parachutes to come
back down to Earth after their celestial victory.

Actually, what all this clearly signals is the end of the
OBJECTIVE, of what is in front of us: this summit, this peak to
be reached any way possible, this object of desire. The flipside,
of course, is the paradoxical rise of the pure TRAJECTORY
involved in an art of displacement in which the very point
of the mountain trek is lost and all that matters is the acceleration of a closed-circuit course, based on what happens in
the stadium. In that static vehicle for Olympic Games which
leads nowhere if not to the podium, the athletes are only ever

convulsionaries, obsessed with excessive records and prey to all kinds of doping, including muscle-building, thanks to intensive training or chemotherapy; subjects of experimentation on the body that no longer even speaks its name to televiewers now addicted to a meaningless progressivism.

We can rest assured of one thing, though if, as our dramatists would have it, language no longer says anything, the same will also prove true of body movement, this art of 'objective' displacement embodied by dance well before it was formalised in the theatre of Antiquity. More than that: if the 'postmodern' era prefers the corporeal to the verbal, the same will shortly prove true of all significant non-verbal behaviour or communication, since the nervous vibrations of interactive synchronisation won't even leave us the time to jig about. It will thereby usher us into the realm of the photosensitive inertia that was once the exclusive secret of the plant kingdom, whereas the animal kingdom seemed to obey the laws of physical movement.

And so, if choreography is a body language, this language will in turn disappear. Such an outcome is already forseeable in the decline in direct communication and the sudden spread of the instantaneous telecommunications whose teleobjectivity outstrips all customary objectivity – of gesture as of word.

What would then remain is the ironic possibility of a caricature, a sort of HIEROGLYPH of human movement, in which the pantomime and the grimace would take over from the smile and the beauty of gesture. For expressionism adapts perfectly to the emerging of these transgenic biotechnologies that teratology and its monsters won't be able to scare off – or hold back – for long, in spite of the alarm bells being rung by cosmetic surgery and the beginnings of a man–machine hybridisation delivered by the nanotechnologies, with the corporal mutation of the transplant revolution taking over from the general mobilisation of the age, already long gone, of the transport revolution . . .

Objectivity, subjectivity. Those two essential philosophical notions will need to be supplemented, one day, by the trajectivity of waves and their instantaneous vibrations. For it is this that will condition the activity, and the interactive reactivity, of a subject who has become a victim of the *domestic terror* that Hannah Arendt saw as the realisation of the law of movement.

To convince ourselves of this, all we have to do, yet again, is look at the recent police use of surveillance targeting ordinary nonpolitical prisoners. Being rigged up with 'electronic handcuffs'[5] is actually a very interesting way of carrying out a sentence, since, we are told, not only does it not cut a felon off from family or social and professional life, 'it actually makes *the prisoner responsible for managing his own punishment'*.

You can just imagine this new SELF-SERVICE serving an individualism delivered to the masses via the TRANSPONDER and via the cell phone of a mobiviewer for whom such a device is now not so much a telephone handset, a *combiné téléphonique*, as the combinative tool of some policy or, more exactly, of a 'proximity market' – a sort of 'flashlight' to dazzle yourself with and, if possible, to dazzle others.

Since television via portable phone is primed to replace the fixed television set, that box that brings together, at home, members of the same community, of the same family, there can be no doubt that, tomorrow, the promoters of mass individualism will justify this now systematic delocalisation – as charity organisations already do, in relation to the panic over clandestine immigration – by asserting the needs of the exiles of a MASS OUTSOURCING that they otherwise exalt in the name of the benefits of international economic integration!

The only question left hanging remains the question of the programme content of this kind of LOCOMOBILE entertainment, whereby the televiewer is often embedded and carted away by a series of AUTOMOBILE carriers, with metro trains and panoramic lifts already having their screens, as though the video control room had taken over from Plato's cave.

'*'He refuses to be alone. He is the man of the crowd'*, observed Edgar Allan Poe, some little time ago . . . But in this conflict of interpretations, the business of the appearance of the real environment is under threat, and it is to be feared that in the near future the photosensitive inertia of the passerby will win hands down over what still remains to him of the customary body language involved in his ambling and rambling. This is already the case with the OTAKU culture of victims of internet addiction.

Unless – as was optimistically claimed by the sentencing judge quoted above in relation to prisoners fitted out with Transponder handcuffs – this new type of 'televisiophonic wire for tying hands' actually does make the wearer more responsible. After all, he has to manage the sentence he has incurred *himself*: he has either to look carefully straight ahead OBJECTIVELY to avoid obstacles, or avert his eyes to acutely contemplate the tiny screen of this clairvoyant flashlight that encourages strabism – divergent or convergent, no one knows which – in a LOCOMOBILE televiewer whose vision of the world's 'relief' was once guaranteed by binocular vision, to ensure his optimal balance and help him avoid falls!

These are all so many illustrations of the vertigo the mind experiences when faced with 'the force of expansion of the evil locked inside the grey rectangle of a screen where shadows that talk go back and forth' dreaded by the Italian writer Guido Ceronetti. Ceronetti went so far as to counsel us all to give up television or, more exactly, to get rid of this receiver that used to be a 'household' appliance once, but is now in the process of sudden miniaturisation, slipping to the very bottom of our pockets; the pulsation of its buzzer, or the little night music of its ring, taking on the task of waking us up – in time or at the wrong time – from our siesta from reality.

Last century is actually riddled with numerous symptoms of such DROMOMANIACAL desertion. Here, again, contemporary art has served to expose the 'gesticulatory trance' of plastic expressivity – in particular with lyrical abstraction, which was opposed to abstract formalism and was also most likely behind Action Painting. That was thanks to Hans Hartung, the legionnaire who had a leg amputated in the Second World War, thereby forming a perfect duo with Blaise Cendrars, the man who had an arm amputated in the First World War, and who brought to literature the ambulatory violence of *The Prose of the Trans-Siberian*.

Master of an arm movement that he executed from his wheelchair, Hartung accomplished a polygraphic ritual in which, he reckoned, 'first and foremost it's the time it takes to execute a line, the slowing down, the speeding up, the misplaced time that counts'.

Whether squirt or scratch, everything that was to extend into the dripping of a Pollock is already there, as though to oppose the geometric abstraction of Hartung's contemporaries, attached as they were to this fixity, this inertia whose model was the icon, as in Poliakoff, for example.

'I don't paint forms, I paint forces', the master of a lyrical abstraction that inherited the violence of the *Blitzkrieg* would ultimately observe. A strange mutation, where the hand of a mutilated man takes over from the legs of choreography; where the trance of the trace left by Dripping prefigures the TRACEABILITY offered by the products of a super-swift age in which the trajectography provided by cruise missile radar monitors seems to be merely a strategic perpetuation of Action Painting!

In one of his courses at the Collège de France, Roland Barthes praised the vibratory energy of the billiard player, of his hesitant yet extremely adroit gesture. But here, the exact opposite happens, since the thinking behind an advantageous tremor is the opposite of this wild fit of anger where reflex dominates all reflection. That is precisely Pollock's contribution to painting, as opposed to Rothko's. Hartung was opposed to Poliakoff in the same way.

Declamatory gesticulation, ambulatory madness – these are so many cabalistic signs of a hidden language, betraying both the decline of the verbal in favour of the visual and the gestural, but, even more, the decline of all representation, since the real time of synchronised interaction was once and for all to replace the time, as well as the space, of action – action that is all too real and enduring.

So after the *automatic writing* and the 'magnetic fields' of Surrealism, we saw the emergence of an *automatic culture* based entirely on the 'just-in-time' delivery of electromagnetic waves that convey our communications and that end, fatally, in the 'zero stock' of representation – not only 'artistic' representation, but political and juridical representation as well. As is abundantly proved by the business of 'author's rights' and 'global licencing' on the internet . . . following the legal case involving contract workers in the performing arts in France.

'Synonymous with adaptation to international economic integration and to the unification of markets, geographic mobility, as Europeans are now aware, can improve your chances of employment', we read recently in a major weekly. At the very same moment we discovered, under the humorous heading, 'The trip is a trap for recruitment', that the unemployed and people living on the poverty line with no job security are the first to be penalized precisely by the requirement of mobility for jobs that demand longer and longer hours to compensate for these distances that never cease growing between people's homes and their workplaces.

So casual workers in the performing arts are now joined by people working on and off as casuals under fixed-term contracts, not just in France but more or less all over the world. This is a development that only anticipates the coming outsourcing to the past that won't fail to occur . . . Yet, even as these problems of mobility operate like so many multipliers of inequalities, the world of transnational corporations is busy organising general mobilisation, thereby running away from the intrinsic dangers of relocation.

The general mobilisation of immigrants of every stripe betrays the end of the old 'geopolitics' of production and of employment, as well as of the fixed abode, in the face of the rise of an emergency 'chronopolitics' that can only end in social chaos; the time of the utopias of the nineteenth century is now giving way to the age of a uchrony characteristic of the history of globalisation.

We can better understand, then, the urgent imperative of prisoners' electronic handcuffs, of this Transponder that is, in the end, merely the double of the cell phone; we can also better understand the imperious necessity for some sort of tracking of the exiles streaming everywhere you look, for an eccentric globalisation in which networking auto-matically involves the perpetual wandering of such exiles, but especially a constant TRAJECTOGRAPHY of their displacements along with the electronic feats of the famous 'Radio-Frequency Identification chips' (RFID).

Another practical consequence of societies' going on-line and networking: if the attaining of autonomy by individu-als goes hand in hand with the acceleration of reality, then SYNCHRONISATION becomes the main threat to civil liberties, since network expansion remains perfectly dubious insofar as its INTERACTIVE structure smacks of an order whose movements have to be spontaneously correlated.

Whence the resettlement of a 'community of interest' tending to the oligarchical and pyramidal in this community of emotion, which is anarchical but radio-controlled, and promoted not, as previously, by the standardisation of public opinion, but indeed by the synchronisation of private emo-tions, collectivised as these are by the interactive instantaneity of the real time of telecommunications and not, now, by the endurance and persistence of the 'real space' of a domiciliary

staying-put which saw the settling of the continent by populations that went from being nomadic to being sedentary.

'Public opinion is a phony *subjectivity* that allows itself to
become detached from the person and put into circulation',
wrote Karl Kraus. As for public emotion in the age of the sychro-
nisation of sensations, that is just a phony objectivity that fools
the individual through (realist) acceleration of his circulation!

The One and the Manifold then surrender their preroga-
tives in the face of the fixed and the mobile. It then becomes
imperative to democratically address the ethical and political consequences of this and to follow up analysis of the role
of the philosophical notions of object (and objectivity) and
subject (and subjectivity) with analysis of the notion of
TRAJECTORY, along with this just-in-time trajectivity, in
the realms of strategic and economic networks, whose consequences are shaking up the political and juridical activity of
nations from top to bottom, just as they shook up the realms
of literature and art, last century. We saw this with Surrealism,
but even before, and especially, in the nineteenth century,
with the invention of the cinematographic parade of the
'seventh art'. If we fail to do so, the globalisation of the Single
Market could well lead to these 'social epilectic fits' that won't
so much be 'revolutions' anymore as repeat catastrophes, even
more dreadful than the ones produced by Soviet Futurism
and Italian Fascism, to say nothing of the Nazism of Hitler's
TOTALE MOBILMACHUNG and its *Blitzkrieg*.

A disenchanted inhabitant not so much of some virtual
CYBERWORLD as of an all too real NANOWORLD,
the mass 'tourist', born of the transport revolution, is scarcely
more now than an eternal return (on investment).

In this tourism of desolation, as in the technologies of the
transmission revolution, the various economic and media
powers are starting to act just like these scientists who are
already able to manipulate atoms, one by one, as though

the world was made of pieces of some micro-Lego that they could put together here or there, as their desire for economic self-sufficiency dictated. The only difference is that, here, the ATOM is the individual and his QUANTUM OF ACTION, the interactivity of a former 'actor' who has morphed into a 'foreign interactor' in a parody of freedom that transforms us, both one by one and all together, into this INTERACTIVE TRIBE we have become. Such a tribe no longer has anything in common with the 'social body' of democracies, but a great deal in common with the social body of Hobbes's Leviathan; with these convulsionaries suffering from a sort of epileptic fit, in which the old connective tissue of empathies (social, familial . . .) is replaced by the disjunctive chaos of antipathy and repulsion towards the other – any other.

And so, the programmed outsourcing of the postindustrial tourist diaspora tends to chime in with the internalisation of a science of the infinitely close and the infinitely deep. In that science, the MEGASPHERE of astronomy's infinitely distant gives way to this MICROSPHERE of the unseen proximity of nanometric matter that disqualifies the 'naked eye' – in other words, insight – just as naked man was once disqualified, that HOMO SACER, sacred man, at the origins of slavery in the Latin world.

This loss of sight of objectivity *de visu* exclusively promotes tele-objectivity in the infinitely vast of televisual telescopy – but every bit as much in the infinitely miniscule of the tunnel-effect miscroscopes used in scanning microscopy. This particular tunnel no longer opens onto the light of some 'emergency exit' anymore. It opens onto the entrance of a triumphal explosion, the explosion of the 'astronomical' BIG BANG, preceding the BIG CRUNCH, that fatal implosion foreshadowed today by the nanometric quest for an INFRA-THIN, the *inframince* so dear to that apostle of extra-retinal art, Marcel Duchamp.

But to conclude, let's go back to our tourists of desolation, these sedentary folk of a new kind, made anxious by far horizons. These DROMOMANIACS of finiteness no longer travel to see the world so much as to witness, live, the scale of a planetary confinement that attracts them the way the edge of a cliff attracts the curious; for these luggage-less travellers closely resemble those extreme sports practitioners who seek strong sensations in a desperate bid to compensate for the inertia of a reality decidedly too down-to-earth.

Similarly, certain travel agents specialising in 'war tourism' offer their clients a new kind of destination: after the trip through the battlefields of the past and the re-enactment of old battles (you might remember the commemoration, in 2005, of Napoleon's victory at Austerlitz as well as that of the defeat at Trafalgar), these agents charter planes so tourists can visit the places where clashes have occurred in the Intifada in Palestine. They even add the economic collapse of Latin America to their tour operators' programmes, after desolated Africa . . . Such *voyeurism* cunningly completes the *exhibitionism* on show in the repeat atrocities relayed by television.

Others go off to Quito to visit the famous 'Volcano Alley', the thirty or so volcanoes, nine of them active, that form the Andes cordillera, running along the Equator. Some time in the not-too-distant future, no doubt, people will be able to scare themselves silly by taking a trip to inspect the concrete burial mounds which cover the sarcophagus of Russia's nuclear power stations, that other chain of volcanoes in constant radioactivity . . . For the devil's envoys, our *visiteurs du soir*, the excitement of an EXOTIC risk takes from the discovery of the foreign customs of an ETHNIC ART now only to be found in the museum.

Since the dramatic clearly now takes the place of the exotic, people are looking to the ends of the Earth to provide those strong sensations long ago initiated by the tragedy of ancient theatre or by, say, the *foire du Trône*, last century's annual Paris funfair, with its high-flying merry-go-rounds.

This is because the unknown now sought is not so much the other, the different, anymore as the enemy of the human race: violent death, in other words the appeal of the end of an era in which the urge to commit suicide is not just psychological anymore, but also sociological and political – as this other, TERRORIST, tourism of the kamikaze would seem to indicate.

This is also what it is, this ambulatory madness of the DROMOMANIAC, whose sole 'hand luggage' is a mere belt or an explosive shoe whereby what is desperately attempted is less some kind of struggle than the final desertion.

So the war tourism discussed above has now been replaced by the tourism of a 'finalistic terrorism' in which terror and the round-the-world trip are one and the same: at once occurring live on the ground and as a recording, thanks to the replay and 'looping' of terrorist attacks; disasters of an era where, in the words of the theologian Dietrich Bonhoeffer, 'we have transformed escape into triumph'.

In another example of such dromomania, a travel agent in California is now offering clients a chance to discover how illegal Mexican immigrants live. Going by the name Reality Tour, Bike Aid, an agent of the non-profit organisation Global Exchange, organises excursions known as Reality Tours around generally dramatic themes. Over a fortnight, these FUGITIVES discover the desert and its landscapes, but also the electronic border between North American and Mexico, with its captors, its night vision cameras and its steel fence soon to be extended over 1200 kilometres by a real 'Desert Wall'.

A journey of initiation, sure, but to learn what lesson? If not the lesson of this FORECLOSURE which the astrophysicist Stephen Hawking talked about in a BBC interview in 2006: 'Sooner or later, disasters such as an asteroid collision or nuclear war could wipe us all out. But once we spread into space and establish independent colonies, our future should be safe.'

So, if in the near future the life-giving star was no more than a mere technical platform, the INTERACTIVE PLANISPHERE desired by the promoters of Progress, it would then remain to write these words somewhere in the desert of the world: STOP EJECT!

Chapter 6

Understanding was once the art of arts. It no longer suffices,
we need to divine.

Baltasar Gracián

Today the most terrible of government duties surely has to
be the duty of lying through deterrence. Such a lie resurrects
the lying by omission that Hannah Arendt so cleverly divined
when she wrote: 'We might not be capable of understanding
ever again, that is, of thinking and expressing things that we
are yet capable of doing.'[1]

Faced with this sudden degeneration in *savoir-faire*, the
universality of the contemporary disaster necessarily implies
immediate radical reform of the University. Such reform
would include setting up a sort of general hospital of science
and its technologies which would aim, if possible, to face up
to this accident in knowledge resulting, not so much from
mistakes and failures, as from the spectacular success of the
technosciences of matter and of life, of which the biotechnologies are the pinnacle, with the instrumental contribution
of the nanotechnologies of the living.

A hospital, then, or rather a hospice of science, a *hôtel des
Invalides* of learning and knowledge once and for all deprived,
it would seem, of a critical conscience, not through their
incapacity to make rapid progress but through an unprecedented success, that no one can actually any longer control,
despite the much touted democratic will.

Since last century, in fact, the will to power of the

technosciences has seriously damaged, if not invalidated, the whole set of acquired knowledge, thereby making the different disciplines suspect, considered as they now are to be unfit for life and for the environment.

Arrogant to the point of insanity, BIG SCIENCE has become powerless to check the excess of its success. This is not so much because of any lack of knowledge as because of the outrageousness, the sheer *hubris* of a headlong rush without the slightest concern for covering the rear; its incredible ethical and philosophical deficit.

Volunteers in the 'total war' of the twentieth century, in a progressive militarisation against the totalitarianisms, the sciences have thus drifted, one after the other, only to wind up much the same as 'extreme sports'. The temptation of theoretical doping has finally won out over the wisdom of a perfectly relative knowledge, in the name of some allegedly Deterrent Peace between the nations involved in a terrorist balance, which the current nuclear proliferation vainly confirms.

'A little knowledge is a dangerous thing', as the saying goes. Is this the case with physics and any other branch of scientific knowledge? That is indeed the question of FALLIBILITY once posed by Karl Popper.

Can one repent of scientific knowledge? Following the Catholicism of the Christian religion, should BIG SCIENCE repent of its deadly excesses as a universalism, repent of the 'damage done by progress'? Faced with the sudden ecological revelation of an accident in knowledge that extends the accident in substances, shouldn't the scientific community, in turn, do its *mea culpa*? Following what happened in the Catholic Church in the twentieth century under Karol Wojtyla, its pontiff, shouldn't this retrospective soul searching, which clearly started with Julius Robert Oppenheimer after the success of the 'Trinity Test', and especially after the bombing of Japan, be revived?

This is where the paradoxical project of radically reforming the university comes in, using as an excuse the failure of the growing success of BIG SCIENCE. It would no longer be a matter of some scientific luminary's litany of repentance, or of a series of Nobels in Physics surrounded by a few Peace Nobels. It would mean official inauguration of this UNIVERSITY OF DISASTER, which would constitute the indispensable *MEA CULPA* now essential to the credibility of a knowledge in the throes of becoming completely suicidal.

In an important essay Karl Popper observed:

> We could say that the origin and evolution of knowledge coincide with the origin and evolution of life and that they are thus strictly linked to knowledge of the Earth. The theory of evolution connects knowledge and thereby ourselves to the cosmos. For that reason, the problem of knowledge becomes a cosmological problem. In days gone by, I already talked about the fascinating nature of the 'problem of cosmology', characterising it as follows: 'This is the problem that consists in understanding the world, including ourselves and our knowledge as a part of that world'. That is how I still see the framework for the theory of the evolution of knowledge.'[2]

It's bizarre that such a blindingly obvious 'logicocosmic' fact has never worried our *savants* in relation to the ECOLOGICO-COSMIC becoming of all practical knowledge, on Earth and nowhere else.

It's true, as Maurice Merleau-Ponty once remarked, that 'science assumes perceptual faith but does not illuminate it', at least not until that not so distant day when 'science will reintroduce little by little what it initially dismissed as subjective. Then the World will close in on itself and, except for what in us thinks and does science, through this impartial observer that lives in us, we will be a part of and a moment in the great object'.[3]

We're there, it's happened. Whence the urgency of a *University of Disaster* already on the drawing board, a university which will ensure that formal logic, dialectical logic and paradoxical logic no longer let themselves be distracted by the pranks played by the fuzzy logics of postmodernity.

According to Aristotle, there is no science of the accident but only of substances, which implies the improbability of any true accidentology. But today it is all too obvious that the accident in knowledge is nothing other than the accident in this 'science of substances' that now winds up accidentally denting all practical knowledge, due to the excessive, the spectacular success of which we are both witnesses and victims.

Over the twentieth century, the militarisation of science made *knowledge* suspect, just as the development of the 'sensationalist' industrial press made *public information* suspect in the nineteenth century. It's now the turn of 'real-time' *information technology* to be on the nose. The decisive question is: what might this discipline created by a new wisdom, this science of the integral accident of the Progress in knowledge, look like tomorrow? Will the same be true of this paradoxical discipline, as is true of justice according to legal expert, Robert Badinter? Badinter claims that 'justice derives from a technique but remains a virtue', and in so saying, reopens the Trial of Galileo; for, although Galileo's sentence was clearly disgraceful, the proceedings were not!

For want of such a re-evaluation, what would then disappear is science, a BIG SCIENCE that would have brought off its own suicide, like some kamikaze of Progress at all costs.

'If you can do something complicated, you can do something simple.' This popular maxim is manifestly false when it comes to the contemporary accident of a BIG SCIENCE so oblivious to its ravages. And the success of this 'epic' science is quite incapable of protecting us against its fatal fallout, as we can see with current debates over burying radioactive waste that will remain toxic, it would appear, for over 200,000 years . . .

So we note yet again that the contemporary accident of the progress in BIG SCIENCE is an accident in uchronic time and therefore an accident in any serious, salutatory long-term planning, the time limits for prevention exceeding the competence of any political authority.

That, too, is what this 'integral Accident' in knowledge is: an accident that integrates an incalculable number of deeds and misdeeds which are unpredictable over the long term, not only in the domain of atomic physics, but also in the domains of biology and population genetics . . .

Faced with the millenarianist expectation horizon of such a catastrophe in thinking, the necessity of a University of Disaster Accomplished ought to entail the necessity of a more democratic approach to research and to scientific and technological development, with the sharing of different kinds of knowledge being capped by the predominant issue of priorities when it comes to economic investment.

What is the POWER OF RISK, in fact, in these very early days of the third millennium? Such technoscientific power hides behind the exalted promotion of a spectacular technology that even boasts of recreating life, thanks to the artifices of a synthetic biology.

In this area, we might for instance note a recent development in the culture of major risks, with a report by the Commissariat général au Plan entitled, 'How can the state deal with insuring new risks?' The report was put out in the spring of 2005 and its initial practical results were felt in 2006. According to this 'exercise in futurology', the asbestos file, with 100,000 potential deaths and 14 billion euros in victim compensation, spells an escalation in collateral damage due to the hasty introduction of multiple scientific and technological procedures (nuclear, biological, and so on), with the phenomenon of the terrorist risk now further complicating the situation.

Faced with such inflation in uncertainties and costs that are hard to 'mutualise' – to spread between various economic agents – insurance systems can now refuse to provide insurance cover on certain risks since, through lack of judgement and prudence, the very existence of the big insurance and especially reinsurance companies would find itself definitively threatened, this time round . . . Whence the urgency, according to the authors of the report, of seriously beefing up the risk appetite of a government, the French State, that hates anticipating catastrophes and even tries to turn optimism into a civic duty!

An OECD report published a few months later, in the autumn of 2005, announced that American insurance companies could no longer cover MEGARISKS alone, in particular those linked to mass terrorism, by definition unpredictable, and so rule out any actuarial calculation likely to fix some kind of 'market price'. Firms were called upon to rethink their type of economic cover for natural or human disasters in this, our twenty-first century.

Finally, in the spring of 2006, the United Nation's World Food Program and the French insurer AXA-RE finalised a contract to insure Ethiopia – the second most populated country in Africa, with 77 million inhabitants – against drought; and they did this, using the 'climate derivative' model which, in rich temperate countries, protects winter sports resorts against lack of snowfall, or winemakers against excess rainfall . . .

This is the first insurance contract in the world to deal with emergency humanitarian aid, and it anticipates those that will shortly prevail for any number of countries that are victims of global warming. The contract specifies, furthermore, that, if the WFP experiment succeeds, the UN and the World Bank could extend the rescue mechanism to Kenya, Malawi and Nigeria.

Faced with such an urgent geopolitical situation and awaiting the foundation of a University of Disaster Accomplished or an Accident Museum, the Ecole des Mines in Paris has just created a doctorate in 'the science and engineering of high-risk activities'. Twenty or so theses on the theme were to be submitted at the end of 2006.

Even if it seems at least a little premature to apply the term 'ENGINEERING' to natural disasters, this does have the merit of opening up the question of a practical collective intelligence of shared interests in the face of megarisks; and it does so at the precise moment when the catastrophe of a fatal convergence of all risks is on the agenda for a BIG SCIENCE hijacked by the accredited experts of Progress at all costs, whose grab for power has wiped out all judgement, all rationality, turning research and development into a culpable passion more than a capable science; a sort of postmodern expressionism destined for the theatre of an epic science that would replay without letup the tragedy of the Garden of Eden, at the foot of the Tree of Knowledge – this time at the risk of expelling shameful Humanity, not from our 'original paradise' anymore, but from Mother Earth, and sending us packing in the direction of the exoplanets so dear to Stephen Hawking.

> The thing, the World, dominates the human being; he is a prisoner, a slave of the World, his domination is merely an illusion, for it is technology that is the force by which the Earth seizes a being and subdues him. And it is because we no longer dominate that we lose our footing, that the Earth is no longer our own Earth. That is why we become more and more foreign to the Earth.

So wrote Dietrich Bonhoeffer as the 1930s got under way. The 'grand illusion' described here is materialising today in the false dromoscopic perspective of an accelerated reality, whereby 'it is a matter of deceiving man, who is effectively in flight, by making him think that this flight is his triumph and that the whole world is in flight ahead of him',[4] concludes the theologian, martyr of the *Blitzkrieg*.

Today, still, the deceptive optical illusion of some sort of Progress, which even touts itself as an ENHANCED REALITY, is merely the accelerated reality of the disaster of a 'free escape', beyond this too small telluric planet, in a desperate search for a world that could possibly replace the fallen world, now reduced to nothing, or as good as, by the force of a technology that has estranged us from geophysics. By that very fact, humanity in its entirety has been dragged into this colonial and meta-geophysical quest for the very latest of New Worlds.

The ultimate in relocation, such global OUTSOURCING would be just one more madness on the part of our progressive civilisations – a madness that would bring off, to perfection, the romantic prophecy: 'If it could be done, we'd give up orbiting the sun and tear off beyond its limits.' That's now a done deal, with these space probes that venture 'into the night of the unknown, or the chill of another world, it hardly matters which, as long as it's foreign to us'.[5]

So it's definitely here and now, on this planet – an endotic planet, not at all an exotic one like the extrasolar planets sought by Corot, the space telescope launched at Baïkonour in December 2006 – that the issue of the day arises and reveals itself to the postindustrial mind. This particular day is not the one shaped by the revolution of the Earth around the Sun, nor is it the alternation of night and day which we had a right to calculate. It is something that determines the very essence of the world along with our existence.

'The Day constitutes the great Rhythm, the natural dialectic of creation, and this rhythm is both rest and movement.'[6] This youngest Day is the REVELATION that makes each of us, through birth and humility, a 'revelationary' more than a 'revolutionary' figure in the great movement of a restless and purposeless progress, which has turned history into a sphere rather than a vicious circle, a dromosphere of accelerating reality. In this acceleration, headlong flight is never more than a triumph without a future; a Pyrric victory in which the avant-gardes' flight from reason is merely a dead end in a will to power whose arrogance seems derisory. Which is now borne out by the current state of the quest for 'extra-terrestrial life', with the abandonment of US government funding for the Search for Extra-Terrestrial Intelligence (Seti) programme and its underlying postulate: 'the banality principle' that pinned its hopes on the discovery of 'signs of life' – since our old planet was so ordinary, so utterly banal, there simply had to be other similar ones out there on which 'life would have emerged just as automatically as on ours'.

This is a strange turn for cosmic evolutionism to take, since it seems our tiny planet is not, in the end, as banal as all that. That fact has hit one exobiologist hard: he now thinks that 'we might well be absolutely alone in the Universe'.

In no way disarmed, even so, by the probable nonexistence of another habitable celestial object, our demiurges have gone so far as to actually suggest the creation of a 'quasi-terrestrial' atmosphere, designed to make a star hostile to life viable. Such 'creationism' goes by the name of TERRA FORMATION . . .

All this is too close for comfort to the crazy excesses of the old Soviet Union and not at all reassuring about the future of a progressivist intelligence or of a UNIVERSAL knowledge, that has turned 'extra-rational' every bit as much as 'extra-terrestrial' – along the lines of the red star of Sputnik, or of those gigantic nature-transforming public works which led populations not so much towards the 'radiant future' of some kind of atheist communism as towards the negative horizon of a COSMISM doomed to degenerate into a 'progressive' cult, a state religion. A marvel of technical know-how, 'creationism' is clearly back. As a materalist cult, it won't fail to go beyond positivism to become a true ILLUMINISM, tomorrow, just as it was in the Soviet Union, yesterday. This belated and ultimately 'extra-solar' sun worship is already foreshadowed by the instantaneity of these airwaves, these mass media that literally mesmerise the mobs.

As Dr Theodore Van Karman, an 'engineer of living systems' recently explained: 'Scientists discover things that exist, but engineers create things that don't exist.'

That just about sums it up: after creating an extra-terrestrial world, thanks to geo-engineering, we could, for example, loop the accelerating sphere of reality, the dromosphere, back on itself, *by creating life*! And we could do this thanks to decoding the human genome . . . Miserable lay miracle of an industry of living systems that would thereby take over from the twentieth-century's industrialisation of death.

We should never forget that this accident, this disaster in knowledge, first struck Western culture, the contemporary arts, in a century that was just as merciless on aesthetics as it was on ethics.

In an essay dating, mind, from 1945, Henry Miller noted: 'Artists today are divided. Part of the time and energy of each one of them is devoted to DESTRUCTION, voluntary, intentional, deliberate destruction. For those who want to follow Picasso, the old world can't die fast enough.'[7]

And so, before hitting science and the *arts et métiers* of technical progress head-on, the urge towards self-destruction, denounced by Arthur Koestler, paraphrasing Freud, struck the whole set of *métiers d'art*, the arts and crafts, of Western culture.

'An optimist sees the opportunity in every difficulty', wrote that expert, Winston Churchill . . . It's impossible to analyse the progress in the difficulties at issue without referring to war, to the great conflicts and their endless preparation over the course of the ages. Ever since the applied sciences first kicked off, whether we like it or not, the ARSENAL has in fact been an 'Academy of Disaster'.

From the alchemist's workshop, through the manufacture of projectiles and their molecular explosives, to the laboratories of poison gases or those involved in the MANHATTAN PROJECT, the arsenal has been the powder store of Progress; the hidden face of the *Janus Bifrons* of the Western university, as Galileo, among others, demonstrated when he offered the services of his astronomical telescope to win the naval battle the arsenal of Venice was gearing up for.

Why wouldn't this perverted military intelligence, a by-product of humanism, one day be asked to help preserve us from the integral Accident, rather than exposing us to it more each day? Here the ancestral duel between arm and armour just can't cut it in the face of the excessive power of a mass destruction inaugurated under the auspices of the military-industrial complex though reviled, in the middle of the Cold War, by President Eisenhower.

In his farewell address, the American general, a specialist in logistics, warned democratic nations against the persistence of this famous complex, which threatened them on the virtuous pretext of saving them from totalitarianism.

Half a century later and despite the implosion of the Soviet Union, we have to acknowledge that that particular complex was not only a threat to the free nations, but just as much a threat to science and to every body of knowledge essential to peace, since the megatonic arsenal of the two Cold War adversaries led to the proliferation we all know about, with the 'unbalance of terror' of an uncontrollable, suicidal diaspora . . .

As early as 1948, the philosopher Albert Camus condemned the fact that the latest theoretical advances in science 'have led it to cancel itself out; its practical improvements now threaten the entire world with destruction'. Camus tried to open up a new path. The learned ignorance of a science outrageously militarised by the requirement of nuclear deterrence between nations could be replaced, for instance, by the tragic lucidity of a knowledge open to anxiety and no longer to the immediate dogmatic satisfaction of researchers in quest of the Nobel

in their specialised field. Yet, until now, no one seems to
have thought of analysing this denial, the nihilism of scientific
materialism. No known scientist, with the exception perhaps
of Karl Popper, has yet dared seriously envisage becoming the
kind of 'optimist' described by Churchill, the man who promised his people only 'blood, toil, tears and sweat'.

But let's get back to the FALLIBILISM of that dire period, at
once so arrogant and so quietly desperate. According to Sir Karl
Popper, critic of determinism, the criterion for distinguishing
between a science worthy of the name and dubious intellectual
constructions is the possibility that the scientific statement in
question be refuted by experiment. Whence the notion of the
FALSIFIABILITY, or REFUTABILITY, of knowledge.

You'll say that that is too obvious, banal, or else that
the concrete experiments which have validated theoretical
physics are numerous and conclusive. All this is true; but such
a 'conclusion' also conceals the urgent need for 'life-size'
experiment on the disaster of the success of such experimentation, so blithely partial and limited!

I'd even say that the way they persist in denying the accident in knowledge currently undermining science, as they
only yesterday denied the fatal consequences of Chernobyl, it
looks as though certain scientists might like to experiment, if
not with the end of history (the history of science), then at least
with the end of the history of living beings. The geophysical
finiteness of the living being's terrestrial habitat proving once
and for all the primary validity of a knowledge that was opera-
tional on a scale of one or two towns, yesterday, one or two
continents, today; but that nothing, absolutely nothing, rules
out testing, tomorrow, on a planet-spanning scale.

Here's a question Popper might have asked, as might
Rabelais: 'Is a form of knowledge that exterminates all life
and thereby all consciousness still a TESTABLE science?!!'
Better yet, perhaps, of what purely 'scientific' value is the
knowledge required for the extermination meted out by the

arsenal of various bodies of knowledge produced by a 'materialism' now on the brink of total nihilism?

It's easy to blame it all on Nietzsche, with his badly misinterpreted 'will' to so-called 'power', which allows you, among other things, to avoid seriously examining the accident that befell that particular will in the twentieth century; namely, the full-size experimentation on the FALCIFORM nature of a technological progress that turns around and saps the spirit of humility in any scientist who rejects the role of demiurge.

An exotic accident, linked to a falsification of knowledge resulting not so much from its weakness as from its force of conviction, this power of a vital will, once so generous, suddenly changed in the twentieth century and turned into the devastating hyperpower of a knowledge that paralyses any acknowledgement of Good as of Evil.

Speaking of such extreme risks, acceptable or unacceptable, let's revisit the British astrophysicist Martin Rees: 'If the downside is destruction of the world's population, and the benefit is only to "pure" science, this isn't good enough.' Rees continues in the same vein, illustrating what he is saying in popular terms:

> I staked one thousand dollars on a bet: 'That by the year 2020 an instance of bioerror or bioterror will have killed a million people.' Of course, I fervently hope to lose this bet. [. . .] But of course, no subject is forging ahead faster than biotechnology, and its advances will intensify the risks and enhance their variety.[8]

Outsourced as we are, we have clearly not touched the bottom but the ceiling, the ozone layer, this light glow that divides the sky from the darkness beyond. On the threshhold of the void of outer space, that unknown quantity of the cosmos, the geophysical boundary, attacked on all sides, drags in the boundary of the physics of this living planet now in the process of desertification. A desertification that has reached the point where the *Future Eve* of progressivist savants is no longer a woman, like the first Eve, but *an Earth* that is one of a kind. And this Earth now finds itself as denuded as Adam, the naked man of Genesis who was a victim of the forbidden fruit of the Tree of Knowledge.

In physics, this situation goes by the name of SINGULARITY; whence the astrophysical quest for an eccentric planet that geo-engineering could possibly make habitable and adaptable to the 'future life' of the earthly genus, the genus *humus* – in other words, the human race.

But when Stephen Hawking tells the BBC (in the summer of 2006), that, 'once we spread out into space and establish independent colonies, *our future should be safe*', he in fact denies the state of play of a knowledge singularly damaged in what it has been based on ever since the world began: the geophysics of matter known as Whole Earth.

Hawking takes his contempt even further, moreover, by attributing to the speed of light the role of ensuring the salvation of history, after the abandonment of geography:

> If we used chemical fuel rockets like the Apollo mission to the moon, the journey to the nearest star would take 50,000 years. This is obviously far too long to be practical . . . However, by using matter/antimatter annihilation, we can at least reach just below the speed of light. With that, it would be possible to reach the next star in about six years.

In those revealing words, where science and science fiction literally merge, the astrophysicist, a victim of postmodern illuminism, not only confirms the self-destructive urge of matter in attaining its goal. He unwittingly brings back sun worship – to the point of idolizing the acceleration of reality, this contemporary hubris of a NEOCREATIONISM that the famous BIG BANG seems to have allowed!

In his moral vision, Bossuet – as quoted by Bernhard Groethuysen – distinguished two great orders: 'the Magnitude of Power and the Magnitude of Poverty'. With the spatio-temporal retention due to the compression of real time, this life-size pollution causes the whole planet to pass from a world order of plenary power to the INFRA-WORLD order of a poverty of resources now recognised by a science of ecology that has turned into a simple eschatology . . .

'*I was everything and everything is nothing*', observed Marcus Aurelius, the Stoic Roman emperor. In a Franciscan frame of mind, one could add today: *I, who am nothing, I pity this World, which is Everything.*

If any magnitude of power indeed returns to (atomic) dust, to its *humus*, including the geophysical magnitude of the globe, the science of physics should in turn get back to this humility that till now justified its *savoir-faire*.

Actually, the finiteness of the Magnitude of Nature has reduced to nothing, or next to nothing, the Magnitude *of Culture* in a philosophy of science once acquired within the immense expanse of a territorial body that now finds it harder and harder to support our eccentricities.

Turning round the phrase of the Psalmist in his hymn in praise of God, we could today cry out: '*Shame on Humanity in the non-expanse of its powerlessness!*'[9] Humility is the truth behind physics, as it is of all ethics. As for 'creationist astrophysics', that has been stopped in its tracks by the principle formulated by Hans Jonas, a principle which has nothing in common with the simple principle of precaution, being the principle of responsibility.

In fact, there is a confusing resemblance between contemporary scientific discoveries and colonisation. Someone discovers something important through science, and technology immediately rushes in and colonises the terrain exposed to the light of day; it then abusively exploits it in the name of the Progress of Humanity, before abandoning it completely.

'When you say big colony you also say big navy', declared Jules Michelet. Today we could paraphrase this declaration of principle as follows: 'When you say big (developed) civilisation, you also say BIG SCIENCE'. It has reached the point where the current repentance of the impenitent imperialists of the past should encompass the ravages of maritime technologies – from the ships of the slave-owners right up to the nuclear missile launchers – as well as the ravages of this 'colonial' technoscience, deprived of a conscience and yet responsible for a human environment now reduced to nothing, or as good as, by its reckless progress.

You can't stop progress, they used to say. No, but today it has stopped all by itself, at the edge of the void, of an interplanetary abyss that puts the finishing touches on the finiteness of a geophysical materiality that once, not so long ago, supported our vitality.

Having till now practised the most extreme of reductionisms and exhausted the fullness of biodiversities, we scarcely inhabit anything more now than an unprecedented telluric contraction, that of the continuum of our material environment. Disabled occupants of an apparently painless cataclysm, we find ourselves helpless in the face of this void, this unknown quantity that some would also like to colonise, thanks to the feats of a *Big Astronautics,* whose spaceships will soon be available for some big ULTRAWORLD landing . . . A colonial adventure that is cosmic or, rather, 'tragi-cosmic', fuelled by the hope of some sort of postmodern, but above all post mortem, ressurrection for a humanity freed at last from terrestrial weightiness as well as from the laws of Earth's gravity.

By way of illustration, here's Henry Miller again: 'All aboard on the road to nowhere, from osmosis to cataclysm, nothing but one big, silent, perpetual Motion, standing still at the centre of this crazy dance, moving with the Earth, even if it wobbles, that's what travelling is!'[10]

'We are entering a period of consequences.'[11] And that's an incontestable fact for anyone wanting to avoid 'the crime against the humanities' perpetrated by progress, for anyone hoping to possibly escape the repetition – meaning the pure and simple automation – of crimes committed by last century's sorcerer's apprentices.

This requirement would lead, not to a 'political ecology' exclusively, as is currently claimed, but to the invention of a 'political economy of speed' instead of just of wealth, in a century driven by an acceleration that is now shaking up the accumulation of knowledge from top to bottom, along with the very reality of our active INTERACTIVITIES – and in so doing, bringing about this pollution of Time distances that polishes off the pollution of the substances our environment is made up of.

'In the twentieth century we learnt the atomic nature of the entire material world. In the twenty-first, the challenge will be to understand the arena itself, to probe the deepest nature of space and time.'[12] So Martin Rees went on to write, thereby illustrating the scientist's anxiety over this acceleration of the real whereby disaster goes potentially universal; for this particular UNIVERSALITY calls into question the feasibility, not of the knowledge specific to this or that scientific discipline anymore, but of the whole body of knowledge acquired over the course of History.

Whether we accept it or not, the dramatic revelation of the METARISK of a convergence of accidents and catastrophes, whether climate-related or otherwise, brings us back to the problematic evoked in the eighteenth century by Leibnitz in his *Theodicy*.

Today, though, it's no longer so much a matter of some anthropodicy of the kind Jankélévitch wanted to see, but indeed of a TECHNODICY, which is dragging us against our better judgement into this fatal 'techno–odyssey' of a future space that has no future, devoid as it is of sense as of reason, since the liberating freeing up of the different branches of knowledge brought on by the cognitive sciences and those famous 'software packages' attacks philosophical wisdom right down to its very foundations, with automation causing statistical thought on number to wipe out even the memory of the complex experience of the phenomenology of the living being.

An anecdote might serve to illustrate these apparently 'iconoclastic' words about the future of computer logistics: at the start of the 'revolution in military matters' ten years ago, Pentagon strategists were already calling this cybernetic process 'the knowledge war'.

After the resounding fiasco of the failure of human intelligence in the New York terrorist attack of 2001, we have no choice but to accept that that particular war is lost, just like the one in Iraq. This is to say nothing of the fallibilism of Tsahal's 'failed raid' in Lebanon, where the term, employed by the Israelis, identified the asymmetrical conflict perfectly as a 'life-size' test, a failed experiment of the kind Karl Popper wanted to see!

Actually, anything that shows *elevation* in a nation's 'magnitude of power' (star wars or cyber wars) makes it vulnerable to the sudden revelation of a 'magnitude of poverty'; and what was manifest in the implosion of the Soviet Union is every bit as obvious for any superpower whatever in 'a period of consequences'!

Apropos the false wit of these erudite demiurges who can tell so much just by looking at the Earth and the sky, but are, on the other hand, quite incapable of telling the time we're living in, let's turn to the Psalmist again: 'Truth shall spring out of the earth; and righteousness shall look down from heaven.'[13] In other words, the truth about the exhaustion of resources will spring out of the earth, and what shall look down from heaven is the righteousness that guides the humble and the just, all those who reject the lyrical illusion of the 'magnitude of power' of an astrophysical science confronted by the enigma of creation – a science for which the speed of expansion of the universe is never anything more than a curtained shutter over the darkroom of the cosmos . . .

Milan Kundera wrote, apropos the literary origins of the novel: 'A magic curtain, woven of legends, hung before the world. Cervantes sent Don Quixote journeying and tore through the curtain. The world opened before the knight-errant in all the comical nakedness of its prose.'[14] The same can be said today of the magic curtain of theoretical astrophysics.

After all, France has just seen the launch of the SUN SYNCHROTRON. This is designed to light up the 'clouds of unknowing' of deep matter, for want of glimpsing, at last, the expanse where the greatness of GOD, THE ALMIGHTY shines in the firmament of his power!

'Without the charm of the unexpected', Kundera goes on to say, 'no character and no great novel would be conceivable. The birth of the art of the novel was linked to the author's sudden self-consciousness, for the novelist is the master of his work. HE IS HIS WORK!'[15] In other words, he is he who is, was and ever will be!

Expect the unexpected, we said at the beginning of this essay, apropos an accident that attacks the entire geophysics of the globe . . . For want of a victory in the knowledge war, we have no choice but to accept that, at the turn of the third millennium, what awaits us is not so much progress in some kind of know-how as the unexpected revelation of the magnitude of poverty of a humility already flagged by ecology, with its secular 'prophecies of doom' flushing out, one by one, the lyrical illusions of the cosmic novel being written by the scholars of science's restless roaming.

V

Revelation

The amputation of accidents doesn't hurt.

Vladimir Jankélévitch

Under the auspices of climate change, the atmosphere has become one long nosocomial sickness whose symptoms never cease to evolve with each fresh medical examination. Little by little, the evidence comes to light. From out of a period of suspicion, we are entering a period of stunning confirmation: the *Big Health* of the climate is gone, volatilised by the greenhouse gas effect of an unassimilable progress in energy.

'In two centuries', notes Jean-Marie Pelt, 'we will have released into the atmosphere practically all the carbon that was fixed in the Earth's entrails, in the form of gas and petrol, for millions of years.'[1] With such a toll, the operative sciences have entered a period of voluntary austerity, of which ecology is merely an early warning sign.

Let's not forget that the term 'austerity' comes from a Greek word meaning 'to burn', 'to set ablaze', and so implies the idea of 'drying up', if not of the imminent plundering by a progress in physics that kicked off, over the summer of 1945, with the Trinity Test heat flash, only to extend, after Hiroshima and especially after Chernobyl, in the evidence – no less explosive – of this delayed-action CLIMATE BOMB recently confirmed by the Intergovernmental Panel on Climate Change (IPCC).

Just a bit further down the track and we'll look on, stunned, at the outbreak of the very first 'civil war' between science and technoscience; whereas we have till now been watching them wage a common 'foreign war' against the human sciences and philosophy.

It is indeed ecology, suddenly so close to eschatology, that opens hostilities today, in this internecine war whose *casus belli* is the accident in a partial knowledge, acquired over the course of the ages but now confronted by the drastic limits of geophysics, for a humanity suddenly reduced to the status of an acquired property.

But let's not forget something else Vladimir Jankélévitch said, which is that, 'with war, man goes on his great metaphysical cure of misery, deliberately rationing himself to atone for a too voluptuous civilisation'.[2] The violence of declared conflicts is never anything but a hypocritical substitute for austerity.

'Austerity', the philosopher goes on to say, 'can sometimes be a penance that modern man inflicts on himself in compensation for, or expiation of, the luxury and countless pleasures that technical civilisation has brought him'.

Note, though, that what is in play today is an austerity of a quite different kind, since the 'civil war' between science and technology, or more exactly between theoretical science and practical technoscience, this time involves an unprecedented mortification, most directly concerning, not the voluptuousness of a 'civilisation based on happiness' anymore, but the whole panoply of satisfactions that we experience in our extravagant consumerism thanks to technological advances. It thereby indirectly undermines the experimental nature of our knowledge, since the probationary scientific experiment now involves nothing less than the globality of the ecosystem – in other words the 'life-size' magnitude of so-called progress in universal culture . . .

Now insisted on, ecological austerity by the same token involves the phantasmal issue of experimentation that is not only *global* but *final,* for a physical science wholly merged with the geophysics of the terrestrial star. We get a hint of this already in *geo-engineering*'s dubious propositions for global cooling through artificial insemination of particles into the upper atmosphere . . . You think you must be dreaming in the face of what is not so much MEGALOMANIA anymore as a sort of MEGALOPSYCHOSIS, on the part of an engineering whose ingenuity has clearly turned into folly or, more precisely, PHILO-FOLLY!

'The dignity of man', Jankélévitch further points out, 'is to participate in reason; his indignity, not to be transformed by such reason: his greatness is misery and his misery is greatness'. From this ensue the two philosophical orders previously indicated: the *magnitude of power* of reason and the *magnitude of poverty* shown by humility.

We should note, though, that such modesty is not the strong point of an extravagantly 'hyperbolic' age that has thrown itself into the weapons stockpiling of an energy race which the philosopher stigmatises: 'Austerity is the watchword in decadent times. In decadence, our own vital rhythms seem, indeed, to match the rhythms of astronomy and the calendar so that the yearly cycle of the seasons makes way for an irreversible degradation.'

Having apparently turned eschatological, the environmental sciences now clearly rest on an official declaration of bankruptcy on the part of a progress in energy whose acceleration threatens, not only the geophysics of the globe, but also the history of humanity along with our probable future. Where the wise labourer of the fable once decided that, 'we might as well work, we might as well go to the trouble; it's the capital that's least lacking', today the reverse occurs: it's the capital that's most lacking.

For an experimental science, such a declaration of bankruptcy is fatal. Throwing it out boils down to venturing into the hazardous domain of a determinist chaos, of a meteorology that no one seriously understands.

This background against which all forms of thought, all theses and their hypotheses, stand out; this context that contains all the texts of scientific experiments, is in the nature of a CONTINUUM which has once again become the major issue for cosmogony, as Martin Rees has recently shown.[3]

Whence the sudden quest for what is missing, for black matter and its missing energy, thought to be responsible for the accelerating expansion of the universe peddled as a notion by astronomers who are, themselves, interested in innovating a sort of METEOROLOGY of the space-time of a COSMOS in which the invisible precipitation of 'clouds of unknowing' might conceal the whole of the matter missing from a universe we only see 5 per cent of, with the stars, the planets and the galaxies that make up our map of the sky. We see so little, in fact, that astrophysicists call this infirmity, linked to the reality effects of gravitational lenses, COSMIC ASTIGMATISM.[4]

And so, after the aesthetics of disappearance and the loss of sight brought on by the NANOSCIENCES and associated microtechnology, we are now seeing a boom in the aesthetics of the (necessary) appearance of the invisible, for a MACROSCIENCE that's gearing up to take over from an astrophysics now stripped of all instruments of perception. Whence the recent launch of an all-out surveillance programme, entrusted to the cameras of the Hubble telescope and aimed at drawing up, if possible, the first three-dimensional map of missing matter.

In a joint declaration, four American personalities who were all high-level architects of the Cold War and of whom Henry Kissinger is one, have just called for global negotiations to denuclearise the planet. The world, in their view, is 'now on the precipice of a new and dangerous nuclear era', with nuclear weapons presenting 'tremendous dangers, but also an historic opportunity'. Accordingly,

US leadership will be required to take the world to the next
stage – to a solid consensus for reversing reliance on nuclear
weapons globally as a vital contribution to preventing their
proliferation into potentially dangerous hands, and ultimately
ending *them as a threat to the world.*'[5]

Published in the *Wall Street Journal* on 4 January 2007 and
two days later in France, this text oddly anticipated remarks
made by President Jacques Chirac to the international press
on 29 January 2007 at his conference in support of 'world
governance of ecology'.

Quizzed by the *New York Times* representative on the
problem of civil nuclear energy in Iran, but also its military
aspect, the French president replied: 'It's not so dangerous
having one nuclear bomb, maybe a second one a bit later . . .
What is dangerous is proliferation.'

Oddly, what was then passed over in silence was the
relationship of cause and effect between the *climate bomb,*
which is about to lead to the relaunching of 'renewable'
civil nuclear energy, and the proliferation of the atom bomb,
which directly proceeds from that, as we all know.

In fact, when Jacques Chirac declares: 'what is dangerous is proliferation', he carefully avoids spelling out that
this particular danger stems from the proliferation of civil
nuclear energy that precedes the proliferation of weapons
of mass destruction.

If the climate alarm sounded this winter by the IPCC were
to contribute, tomorrow, to a systematic relaunch of civil
nuclear energy in the name of some so-called conservation
of the environment, it would contribute every bit as much
to the wildly exploding proliferation of atomic weapons that
constitutes one of the biggest risks in history, in the words of
the American signatories in Henry Kissinger's entourage.

Surely we can't fail to register, here, the profound ambiguity in this fatal coincidence between the ecological imperatives

of conserving the earthly atmosphere against the greenhouse gas effect, produced by fossil fuels that are in the process of being exhausted, and the imperatives of preserving peace, as well as public health, against the devastating effects, not only of nuclear weapons in 'terrorist' hands, but also of radioactive fallout. For this, Henry Kissinger recommends swiftly 'providing the highest possible standards of security for all stocks of weapons, weapons-usable plutonium, and highly enriched uranium everywhere in the world; getting control of the uranium enrichment process' as well as 'halting the production of fissile material for weapons globally; phasing out the use of highly enriched uranium in civil commerce and removing weapons-usable uranium from research facilities around the world and rendering materials safe'.[6]

Aware as we are of the extreme ambivalence underlying the international energy situation, it seems hard today to doubt, not the greenhouse gas effect behind global warming, but rather this strange 'transpolitical' coincidence that now drives the big petrol groups to invest in uranium mines and even to exploit nuclear power stations directly, as the new CEO of Total announced they would be doing at the start of 2007. They were doing this, he said, 'with a view to diversifying the group's activities'.

We might also note that the famous BP Amoco group has just transformed its acronym into Beyond Petroleum! We are a long way here from the urgent recommendation to 'denuclearise the planet', which is just wishful thinking on Henry Kissinger's part. It looks more like a revival!

When they try to tell us that the development of nuclear power will favour the protection of the climate against the ravages of fossil fuels, they leave out the warning sounded by the four former American leaders, who specifiy in their manifesto that, 'Apart from the terrorist threat, unless urgent new actions are taken, the US soon will be compelled to enter a new nuclear era that will be more precarious, *psychologically* disorienting and *economically even more costly than was Cold War deterrence.*[7]

In plain language, after the era ruled by the notion of 'Mutually Assured Destruction' (MAD) and its all-out military deterrence, we will be entering an era of 'cold panic' which will in turn require the costly innovation of some sort of organised civil deterrence; the 'ethologically correct' practice of protection against (environmental) pollution this time demanding not a 'Ministry of Fear,' as last century, but some sort of transnational threat-management body. The project to create a United Nations Environment Organisation should provide a useful model.

We might also note while we're at it that on 8 February 2007, the European Commission tabled a directive aimed at making non-respect for the environment an actionable offence, punishable by law, which would mean that those guilty of GREEN CRIMES could be prosecuted. Apart from the discharging of pollutants into the air, water, fauna or flora, the directive also targeted 'the fabrication, treatment, stocking, use, transportation or importation of nuclear material or any other dangerous "radioactive" substances likely to cause death'. All this, just so the European Community member states could finally put dissuasive penal sanctions in place. In the most serious cases, such sanctions would include prison sentences 'able to give rise to extradition'.

For the greenhouse gas effect to end in 'incarceration' of the guilty - well, it's only logical, ecological even! As for 'extradition', in this instance that would be a kind of externalisation - outsourcing - which should eventually spark philosophical debate over the future of the imminent 'deterritorialisation' of all forms of knowledge!

'Denuclearising the planet' is an aim that exhaustion of fossil fuels could certainly frustrate, but the essential question remains: should we understand by it that we need to relocate BIG SCIENCE, nuclear physics? Move it out of the geophysics of a planet now threatened on all sides by the acceleration of its dazzling successes, ever since the flash produced by the Trinity Test at Los Alamos? . . . If such is indeed the case, then the 'deterioration' of progress ought to end in other types of relocation, for a scientiific and technical enterprise whose toxic discharges and fallout into the terrestrial atmosphere do indeed constitute a delayed-action bomb, clinical symptoms of the long nosocomial sickness mentioned earlier.

Recently an ad appeared in the paper that set up a comparison by posing the question: 'Can we reconcile performance and protection of the environment?' Since the DROMOSPHERE opposes the BIOSPHERE and its living diversity with the full force of conviction of its hi-performance advertising, the answer was this: 'Our strategy, based on the complementarity of water power industries and a clean environment, brings you life's essentials.' The emphasis said it all. Unless this paradoxical situation, unprecedented in the history of knowledge, actually ends in the expulsion, the extradition, of a science without patience and without much time, as though its radiant future had suddenly turned EXTRATERRESTRIAL, just as the astrophysicist Stephen Hawking clearly explained it would.

Today, it is revealing to note that the immoderate use of hi-technologies and the negative utilisation of knowledge acquired last century are finally ending in 'the rape of the crowds by propaganda', just as Serge Tchakhotine described it. But also in the rape of a science no longer deprived only of a conscience by its militarisation, but also of the patience necessary to test its validity with regard to the terrestrial environment.

Validated in space – the short term of successful experimentation – the practical sciences should also be validated, or not, in time – the long term of the consequences of their discoveries. Look, for instance, at the unresolved problem of the toxicity of radioactive waste.

Bound to very long durations in the realm of astrophysics, scientists today seem helpless as soon as it comes to the time involved in geophysics, in its geology. Actually, globalisation is nothing more than a borrowed name for this sphere of impatience belonging to the science of physics involved in the acceleration of a reality whose incontinence leads straight to the inertia of history – the political history of events as well as the scientific history of knowledge.

In less than two centuries, speed has thereby replaced physical expanse to the point of making us lose all magnitude and so all measure, not only in the realm of the nature that harbours us, but every bit as much in the realm of culture. Whence the philosophical urgency of studying a disaster that has no known equivalent, since it is the disaster of perfect achievement, of a spectacular success. Only, a success that gives rise to this warning: 'Any scientific truth is only ever an error in remission due to the relative lack of expanse of the terrestrial star!'[8]

Having finally replaced the REVOLUTION, all the revolutions (including the Copernican one) by this REVELATION, whereby *present time* instantly revives the *bygone time* of the representations of the past, we now need to fathom the ultimate consequences of a speed race that wipes out the very memory of the judgement and wisdom accumulated over the ages. The excessive importance given to live presentation, the revelation of instantaneity, in fact dominates history. Including the history of a 'geophysical' physics outmoded by astrophysics and its extraplanetary delirium, in which the exodus of scientific reality is merely the out-of-this-world exile of knowledge acquired in this world.

We might remember that, in the middle of the crisis in European philosophy, Edmund Husserl wrote: 'The Earth does not move.' Whence the sudden untimely looming of a TRAGI-COMIC science such as Albert Einstein envisaged when he poked his tongue out at the photographers and declared: 'There is no scientific truth!' After Copernicus and Galileo, who are still on trial, after Newton and Einstein, the final say today, then, reverts to Stephen Hawking, the scientist who is not so much reactionary as revelationary. Hawking declared to the BBC that the future of humanity depends on discovery of an exotic planet to colonise, since, 'sooner or later, disasters such as an asteroid collision or nuclear war could wipe us all out'.

The astrophysicist emeritus curiously illustrates such an accident in cosmic traffic in such a way that the collision of celestial objects is just like the pile-up of vehicles involved in the transport revolution, in the crazy excess of the head-long rush of progress. It's a bit as if the CRASH TEST had become the symbol of a CRASH SCIENCE to come, the 'other solution' (if we might call it that) only ever being the eternal return of a COLONIALISM that would this time be based on astronautics, as it once was based on some mari-time power's arsenal like the Venice that inspired Galileo. This is not to leave out the rockets of the military–industrial complex denounced by Eisenhower. It's all as though the DROMOSPHERE had only ever been the 'big vehicle' of Western thought, its energy, at once kinetic and kinematic, hurtling our civilisations towards this big progressive 'mul-tiple crash' in which nature and culture in the end prove incompatible.

We might even come to imagine tomorrow's reality-acceleration sphere as something like a racing-car that goes like a rocket, a FIREBALL, whose stored-up energy, taking mass and inertia into account, represents a potential for self-destruction that we will have to get around to studying one day or another, since the kinetic energy of such a cosmic fireball could be dissipated by an 'ecological' stocktaking of nature's time distances as well as of the substances that shape space. Rejecting such a precaution, on the other hand, would entail an 'eco-systemic' catastrophe, a DROMOSPHERIC cataclysm infinitely more serious than the cataclysm of the

ATMOSPHERIC warming of the globe – due to the very fact of the violent dispersal of energy constituted by the excessive speeds of all kinds in a reality that is not only 'historic' but 'geographic'; the twilight of this world's places leading fatally to the starless night of a BLACK HOLE!

In fact, the long focal length of progress in knowledge is not infinite but limited, whether we like it or not, by the relativist non-expanse produced by the recent spatiotemporal contraction of geophysical distances. Of course, we all know that, as Einstein said, 'speed is not a phenomenon but the relationship between phenomena'. It is a sort of phenomenological ANTIFORM in which relativity is essential – on the condition that we consider the latter *here below*, here and now, in a determinedly *down-to-earth* fashion, and not solely *up above* and *over there*, at the very furthest reaches of a universal expansion perceived through the 'telescopic' instruments that allow us to contemplate the firmament.

We note, furthermore, that the focal length of reality acceleration has just given rise to a project for a new linear accelerator around 35 kilometres long to be built in Beijing. Proposed in February 2006 by an international team of physicists, this long tunnel for accelerating electrons and positrons would be the technical complement to the grand accelerator, the *circular* Large Hadron Collider, built beneath the Franco-Swiss border.

We've studied the line and the circle, so why not start studying, right now, the sphere of acceleration of the reality of a manifest progress that doesn't enlarge, but only shrinks? That progress is, after all, primarily responsible for the pollution of the distances of the life-size globe, this DROMOSPHERE that also happens to harbour all life, all known vitality, in the universe.

Since pollution of nature's substances and (meteorological) warming go hand in hand with pollution of life-size (dromological) distances, the issue of the energy economy would find its most complete ecological basis here, and thereby bring back into play not only the *way of life*, but, on top of that, the *mode of speed* of the history to come.

To stick with a purely 'materialist' approach to the threat of the climate bomb, as people today do, means accepting voluntary blindness and refusing to see, not only the contradiction, but the incompatibility between the ecological imperatives of safeguarding an expensive way of life and the hysteria of a developed society that no longer contents itself with a CONSUMERISM fuelled by the finite products of industry, but further and especially, demands unbridled consumption of the time–space of distances on a planet that is decidedly far too small to host the speed race of living beings far much longer.

Basic models for the ECUMENICAL UNIVERSITY OF DISASTER, the majority of sizeable industrial firms these days endow themselves with outfits specialising in research into the 'passive safety' of traffic; as though the *Blitzkrieg* of Progress at all costs now finally ended in outmoding the old passive defence of the populations of yore, with their shelters and ramparts, those fortifications based on that guru, Vauban.

In these eerie factories where engines of all kinds are destroyed, the engineers in charge of the process assess the results of impacts and the deformation of materials caused by 'car crashes' or train wrecks; in other words, the damage done by the dispersal of kinetic energy stored up in the tools of fast transport.

Yet, far from being an attack 'in action' on the acceleration of the performance of the vehicle in question, these repeat CRASH TESTS don't so much target excess speed as improvement in passenger safety; whence the term 'passive safety', which serves as an additional argument for conducting the tests in the first place.

The SNCF, French Rail, have been at the top of their game ever since the railways took off. With technologies like the BLOCK SYSTEM of traffic regulation and

CAP SIGNAL automation that engines are equipped with today, and with their trains running at 300 km an hour, the SNCF have a clear lead in this domain. Their RUNAWAY TRAIN, the TGV, holds the world speed record at 515km an hour and they have had only a single derailment to date, and even that was without a single victim.

As Dominique Tessier, an SNCF executive, pointed out after a successful attempt to beat the record: 'It is important to us to find the right compromise between energy consumption and time gain, which becomes marginal beyond 300km an hour or so. Over a trajectory of 800 kilometres, 20km an hour will only gain you around ten minutes.'[9] Here, the discrediting of physical expanse is clear, since, as the executive goes on to say, 'It isn't the deciding factor in market share. On the other hand, the possibility of accelerating is essential for traffic regularity since it allows you to compensate for possible delays.'

And so, with the aim of ensuring the best possible 'passive safety' along with regularity of service, those running the SNCF facilities study the most perilous situations, from the classic telescoping, jacknifing and derailment of trains through to collision between a train running at high speed and, say, a nuclear container . . .

A perfect illustration of the proliferation not only of the race for weapons of mass destruction, but equally, and every bit as much, of the race for speed to be free of all territorial expanse, the railway metaphor was used at the very same moment by an Iranian president boasting of his determination to ensure the passive defence of his country by means of nuclear energy, no matter what the cost: 'It's a moving train', he said, 'that no longer has any brakes or reverse gear – we ripped everything out and threw it away last year.'[10]

Now there's a vision of prevention, for you! One based on the quasi-inevitability of a major risk in order to prime us, right now, for the disaster lurking just round the corner!

As a specialist in another area, the field of intensive tele-communications, explains:

> Today innovation is on everyone's lips, but there aren't many people who know how to do it efficiently because they get their priorities wrong; they try to manage technical innovation by, say, creating data bases, whereas you should first of all look after managing *the risk generated by the innovation* – the risk of breakdown, of safety for the user, of nonconformity to legal, ethical or social developments, and so on. *What we should be doing now is moving from innovation management to risk management, by reacting in real time when faced with any dysfunction.*[11]

RISK STUDIES is the science of prevention of risks of all kinds. But it is not an ACCIDENTOLOGY, since it refuses to analyse the very effect of the speed of progress and its acceleration on the mapping of 'azimuthal equidistant', or 'great circle', projections involved in the geography of transport networks; on, say, regional development, or the consequences of the METROPOLISATION of settlement, the relocation of firms or the OUTSOURCING of postindustrial production.

Based from the very beginning on an energy choice that goes back to gunpowder and the steam engine, followed by the 'internal combustion engine', our dromocratic civilisation launched itself into the race for an historic acceleration without any brakes. This race was bound up with the urgent imperative of inordinate growth, in which expenditure of energy gradually became a virtual public defence of society's *train de vie*, its 'lifestyle'; a virtual inevitability, not so much political as trans-political. Its unacknowledged symbol was the invention of the 'jet engine', and its jet pipes, anticipating the extraterrestrial liberation of rockets and the innovation of this speed of escape from terrestrial gravity that leads to nothing but the vacuum of outer space; the TURBOCAPITALISM of the single global market, for its part, ushering in the integral accident in the world-economy that's not too far off now. The symptoms of this particular bomb, delivered by an involuntary CRASH TEST, have followed one after another ever since the famous BIG BANG produced by the interconnectivity of the international stock markets' PROGRAMME TRADING.

Actually, the critical situation humanity now finds itself in further calls into question the ill-considered growth in urbanisation and motorisation and, with it, in energy consumption. It therefore remains the major challenge in the battle now underway against atmospheric pollution and its pandemics.

We might note, for instance, that in certain underprivileged countries of Africa, the demand for transport is increasing even faster than gross domestic product or population growth. It has reached the point where a radical shift in social behaviours to do with transport and its velocity will soon be needed. A radical shift to do with metropolitan development will also be needed, in a bid finally 'to control the nature and length as well as the frequency of our different trajectories', as a talented Third World urbanist hopes is possible.

OBJECT, SUBJECT, TRAJECTORY: the impact of the latter on the other two implies, as of this moment on, an economic and political acceptance of responsibility for a particular form of pollution – this one dromospheric – of the time distances of the geographical expanse involved not only in the physical transportation of people and goods, but, even more so, in the transmission of interactive information on a global scale. Without such acceptance of responsibility, an accident in the DROMOSPHERE of accelerating reality will lead to the fusion of a CRASH TEST of capitalism and a CRASH TEST of an auto-immobilism that is heading straight for disaster – the disaster that the implosive cosmology of BLACK HOLES may well be a premonition of.

But to get back to passive safety again for a moment: passive safety fully embodies the ambiguity of a precautionary principle that refuses, in the end, to turn into the principle of responsibility Hans Jonas hoped to see. Such a principle would

require, not the passivity of the security-conscious ideology
of the promoters of the race for profit and its market shares,
but the activity of a defence equal to the ecological risks run
by humanity. For, today, 'the race for energy reserves has
replaced geographical conquest, and old disputes that used
to be resolved in the din of cannons are now resolved by the
signing of contracts' – now that the real time of acceleration
has once and for all supplanted the expanse of the real space
of nations. So often a willing victim of gigantic machinery –
first mechanical, then electronic – which, 'instead of liberating human beings, restricts their autonomy', the individual,
more atemporary than 'contemporary', gradually becomes a
servant of a panic-stricken ecosystem he depends on to exercise what André Gorz, again, calls 'incapacitating professions'
that exile him from his vitality. This has reached the point
where some are already forecasting that, in the near future,
the university and its faculties could well become, for postindustrial society, what the factory and its mills were, yesterday,
for the nascent industrial society . . .

That is a vision of the future in which the accident
in knowledge could well lead, this time, to the implosion of the progressivist DROMOSPHERE. The effects
of this would be analogous to those of Beijing's proposed
INTERNATIONAL LARGE LINEAR COLLIDER,
which is so closely linked, in its 'tunnel effect', to the effects
so often put to work by the people conducting CRASH
TESTS with the aim now of further accelerating the comfort
of private citizens and individuals, rather than advancing the
physics of elementary particles. The gigantic energy generated by such collisions will finally end up reproducing the
presumed conditions of the BIG BANG after a few mil-
lionths of a second.

The initial cost of this cosmic detonation tunnel is a mere
five billion euros, which is nothing compared to the probable bill for the CRASH TEST provided by the 'dromo-

spheric collider' which is likely to generate – here below – the devastating effects of the BIG CRUNCH of the world economy and its ecosystem!

If, indeed, an ecology of substances comes at a price, and we can clearly see that it does, the ecology of distances, *grey ecology*, is priceless, since its economy is on the verge of being permanently outsourced to the middle of nowhere.

Notes

Intuition

1. Marc Augé, *Le Temps en ruines*, Paris: Galilée, 2003.
2. Søren Kierkegaard, *The Concept of Dread*, translated by Walter Lowrie. Princeton: Princeton University Press, 1968 (original publication 1844).
3. Marc Augé, *Le Temps en ruines*, op. cit.
4. The Musée du Quai-Branly, designed by Jean Nouvel, has now proved the point.
5. 'Les Chemins écartés de l'évolution', an interview with P. Berthommeau and J. Albert, *Sud-Ouest*, 2007.

The Waiting Room

Chapter 1

1. Maurice Merleau-Ponty, 'Eye and Mind', translated by Carleton Dallery in *The Primacy of Perception*, edited by James Edie, Evanston: Northwestern University Press, 1964, revised in *The Merleau-Ponty Aesthetics Reader*, edited by Michael Smith, Evanston: Northwestern University Press,1994, pp. 121-149 (original publication 1964).
2. Maurice Merleau-Ponty, *The Visible and the Invisible*, translated by Alphonso Lingis, Evanston: Northwestern University Press, 1969.
3. Hölderlin, 'Hyperion', in *Hyperion and Selected Poems*, edited and translated by Eric L. Santner, New York: Continuum, 1990.

4. Edwin Chargaff, *Le Feu d'Héraclite*, Paris: Viviane Hamy, 2006.

5. André Gorz, *Letter to D.*, translated by Julie Rose, Cambridge: Polity Press, 2009 (original publication 2006).

6. In certain schools of war military geography is no longer taught.

7. For his part, Gilles Deleuze wrote: 'The eye is not the camera, it is the screen.'

8. Jean-Marie Pelt, *L'Avenir droit dans les yeux*, Paris: Fayard, 2003.

9. *Sud-Ouest*, 2006.

10. *Ibid*.

Chapter 2

1. Adrien Barrot, *L'Enseignement mis à mort*, Paris: Librio, 2000.

2. Henri Michaux, *Passages*, Paris: Gallimard, 1951.

3. *Ibid*.

4. Martin Rees, *Our Final Century*, London: Arrow Books, 2003, pp. 119–120, 121.

5. *Sud-Ouest*, summer, 2006.

6. Cf. 'Espace–temps, et s'il fallait tout reprendre à zéro?', *Science et Vie*, September 2006.

Photosensitive Inertia

Chapter 3

1. Dietrich Bonhoeffer, *Creation and Fall: A Theological Exposition of Genesis 1–3*, edited by John W. de Gruchy, translated by Douglas Stephen Bax, Minnesota: Fortress Press, 1997.

2. Marc Augé, *Les Ruines du temps*, Paris: Galilée, 2005.

3. Thomas L. Friedman, *The World is Flat*, London: Penguin Books, 2006.

4. Danielle Chaperon, *Camille Flammarion: Entre astronomie et littérature*, Paris: Imago, 1997.

5. H. G. Wells, *Star Begotten*, New England: Wesleyan University Press, 2007 (original publication 1937).
6. Paul Virilio, *Polar Inertia*, translated by Patrick Camille, Los Angeles: Sage Publications, 1999 (original publication: *L'Inertie polaire*, Paris: Christian Bourgois, 1990).

Chapter 4

1. John Steinbeck, *Travels with Charley in Search of America*, London: Penguin Books, 1997, pp. 70-73 (original publication 1962).
2. Hannah Arendt, *The Origins of Totalitarianism*, New York: Harcourt, 1979, p. 465 (original publication 1948).
3. Maurice Merleau-Ponty, *The Visible and the Invisible*, op. cit.
4. Joseph Roth, *Flight Without End*, translated by David Le Vay, Beatrice Musgrave, London: Penguin Books, 2002 (original publication 1937).
5. Danielle Chaperon, *Camille Flammarion: Entre astronomie et literature*, op. cit.
6. *Ibid.*
7. Note the German prisoners belonging to the Red Army Fraction and incarcerated at Stanheim.
8. Auguste Rodin, *Art: Conversations With Paul Gsell*, translated by J. De Caso and P.B. Saunders, Wilmington: Quantum Books, 1987 (original publication Paul Gsell, *Rodin. L'art,* Paris: Grasset-Fasquelle, 1911).
9. Thomas L. Friedman, *The World is Flat*, op. cit.

The University of Disaster

Chapter 5

1. *Sud-Ouest*, summer, 2006.
2. *Sud-Ouest*, spring, 2007.
3. Fernando Pessoa, *The Book of Disquiet*, edited and translated by Maraget Jull Corta, New York: Serpent's Tail,

1991 (see also Richard Zenith, London: Penguin Books, 2002).

4. *The Bible*, Psalms, Psalm 150, final doxology.
5. The TRANSPONDER, that object whose political importance I already noted so long ago.

Chapter 6

1. Hannah Arendt, *The Origins of Totalitarianism*, op. cit.
2. Karl Popper, *The Open Universe: An Argument for Indeterminism*, Cambridge: Cambridge University Press, 1988 (original notes date from 1956-7).
3. Maurice Merleau-Ponty, *The Visible and the Invisible*, op. cit.
4. Dietrich Bonhoeffer, *Creation and Fall*, op. cit.
5. Hölderlin, 'Hyperion', op. cit.
6. Dietrich Bonhoeffer, *Creation and Fall*, op. cit.
7. Henry Miller, *The Air-Conditioned Nightmare*, New York: New Directions, 1970 (original publication 1945).
8. Martin Rees, *Our Final Century*, op. cit. p. 125, p. 74.
9. *The Bible*, Psalms, Psalm 150, final doxology.
10. Henry Miller, *The Air-Conditioned Nightmare*, op. cit.
11. Winston Churchill, *c*.1939.
12. Martin Rees, *Our Final Century*, op. cit., p. 148.
13. *The Bible*, Psalms, Psalm 85.
14. Milan Kundera, *The Curtain: An Essay in Seven Parts*, translated by Linda Asher, London: HarperCollins, 2007.
15. *Ibid.*

Revelation

1. 'L'Ecologie n'est pas un groupe politique', interview with P. Verde, *Sud-Ouest*, February 2007.
2. Vladimir Jankélévitch, *L'Austérité et la vie morale*, Paris: Flammarion, 1956.
3. Martin Rees, *Our Final Century*, op. cit.

4. Jean-Marie Augereau, 'La première carte en trois dimensions de la matière noire', *Le Monde*, 9 January 2007.

5. 'Guérir de la folie nucléaire', *Le Monde*, 24 January 2007.

6. *Ibid.*

7. *Ibid.*

8. Cornelius Castoriadis, *Fenêtre sur le chaos*, Paris: Le Seuil, 2007.

9. 'Quelle vitesse le TGV pourrait-il atteindre?', interview with Michel Waintrop, *La Croix*, February 2007.

10. Mahmoud Ahmadinejad, *Le Monde*, February 2007.

11. Martin Illsley, 'Les Entreprises doivent maintenant s'organiser pour entrer dans l'économie du temps réal', *Le Monde*, 27 February 2007.

Index